COASTAL WESTERN INDIA

Books of the XCHR Studies Series

No. 1 : João de Barros: Portuguese Humanist and Historian of Asia/ *C.R. Boxer*

No. 2 : Coastal Western India: Studies from the Portuguese Records/ *M.N. Pearson*

No. 3 : Black Legend of Portuguese India (The): Diogo do Couto, His Contemporaries and the Soldado Prático-A Contribution to the Study of Political Corruption in the Empires of Early Modren Europe/ *George Davison Winius*

No. 4 : Health and Hygience in Colonial Goa (1510-1961)/ *Fátima da Silvà Gracias*

No. 5 : Trade and Finance in Portuguese India: A Study of the Portuguese Country-Trade, 1770-1840/ *Celsa Pinto*

No. 6 : Goa and the Revolt of 1787/ *Charles J. Borges* (Ed.)

No. 7 : Goa and Portugal: Their Cultural Links/ *Charles J. Borges* and *Helmut Feldmann* (Eds.)

No. 8 : Goa-Kanara Portuguese Relations 1498-1763/ *B.S. Shastry* and *Charles J. Borges* (Eds.)

No. 9 : Goa's Foremost Nationalist: José Inácio Candido de Loyola: The Man and His Writings/ *Charles J. Borges* (Ed.)

No. 10 : Goa and Portugal: History and Development/ *Charles J. Borges, Óscar G. Pereira* and *Hannes Stubbe* (Eds.)

No. 11 : Conversions and Citizenry: Goa under Portugal 1510-1610/ *Délio de Mendonca*

About the Author

M.N. Pearson was born in New Zealand in 1941, and has B.A. and M.A. degrees from the University of Auckland. His Ph.D. (1971) is from the University of Michigan. From 1970 to 1975, he taught in the Departments of History and South Asian Regional Studies at the University of Pennsylvania. Since 1975, he has been at the School of History, University of New South Wales, where he is a senior Lecturer. He has spent considerable time in western India, especially in Goa, Bombay and Baroda. Among his other publications are *Merchants and Rulers in Gujarat: The Response to the Portuguese in the Sixteenth Century* (University of California Press, and Munshiram Manoharlal, 1976), and (co-editor) *The Age of Partnership: Europeans in Asia before Dominion* (University Press of Hawaii, 1979). He is currently co-editing, with Ashin Das Gupta, a collection of essays to be called *India and the Indian Ocean, 1500-1800.*

XCHR Studies Series No. 2

Coastal Western India

Studies from the Portuguese Records

M.N. PEARSON

CONCEPT PUBLISHING COMPANY PVT. LTD.
NEW DELHI-110059

ISBN: 978-81-7022-160-9

Edition 2023

Published and Printed by

Concept
Since 1974

Concept Publishing Company Pvt. Ltd.
A/15-16, Commercial Block, Mohan Garden,
New Delhi-110059 (India)
T : 91-11-41101460
E : publishing@conceptpub.com, *W*: www.conceptpub.com

TO
MARGOT

INTRODUCTION

THE Xavier Centre of Historical Research is justly proud of commemorating the second anniversary of its existence with the release of the second monograph of its *XCHR Studies Series* which was inaugurated a year ago with Prof. C.R. Boxer's *João de Barros : Portuguese Humanist and Historian of Asia*.

Dr. M.N. Pearson has already made his mark in Indian historiography with his *Merchants and Rulers in Gujarat* as a trend setter, laying bare private interests and unofficial compromises underneath the diplomatic façade of intransigence which alone had been played aloud by the historians of the Mughal empire or of the western colonial expansion into Gujarat. This present selection of essays continues this welcome trend and extends it to the rest of *Coastal Western India* of the sixteenth and seventeenth centuries. This will stand out as the best effort on the part of a non-Indian historian to do justice to the Indian component of Indo-Portuguese history.

PANJIM,

1 April 1981

DR. TEOTONIO R. DE SOUZA, S.J.
Director
Xavier Centre of Historical Research

PREFACE

THIS BOOK consists of five previously published articles, and two unpublished papers, all of them dealing with aspects of the history of coastal western India in the sixteenth and seventeenth centuries. There are several reasons for my deciding to put together this book. Most obviously the two unpublished papers are now made available, and I hope they contain some material which will be of interest. Some of the published articles appeared in obscure journals or collections, and this book makes them, for what they are worth, more accessible.

More important, I hope this book will contribute to two closely related and fast expanding areas of historical research. These are Indo-Portuguese history, and maritime history. The former is going ahead by leaps and bounds, helped by the establishment of a regular International Seminar of Indo-Portuguese History, and by the recent founding in Goa of the Xavier Centre of Historical Research. The latter field is also at present flourishing, as was seen at the Seminar on Indian Maritime History at Santiniketan, West Bengal, in March 1979, and in August of the same year at the International Conference on Indian Ocean Studies in Perth, Western Australia. Finally, producing this book has not only given me the opportunity to revise some of these pieces, it has also forced me to reflect on the themes they cover, and to consider general problems and future possibilities in these areas of research. This being the case, this introduction will regrettably at times be rather autobiographical. But I feel there is some value in an author trying to make explicit the themes and methods he has found to be of value in his own work.

It is first necessary to make clear the difference between these two fields of maritime and imperial history, for there seems to be some confusion on this score. Some people apparently see them as identical, and implicitly find maritime history tarred with the same brush as imperial history. This attitude no doubt simply reflects the fact that much maritime history has been overly concerned with the military and other activities of Europeans in the Indian Ocean. But this point of view clearly is nonsense. It is debatable whether sea power was even a necessary pre-requisite for European colonial rule in Asia; certainly, and obviously, the actual conquests took place on land. Such a point of view also ignores centuries of Indian maritime activity before da Gama, none of which, except perhaps that of the Colas, can be considered to be "imperial". Finally, several recent studies have demonstrated the continuance, even florescence, of Indian maritime commerce at least until the eighteenth century. The two fields no doubt overlap at times, but are nonetheless clearly distinct.

What appears from the essays in this book, and from the contributions to the various conferences and seminars mentioned above, is that both these fields of research suffer from severe problems of definition. These are particularly acute in their relations to the field of Indian history itself, or what can be called "internal" Indian history. Take first the expanding field of maritime history. An easy definiton of this area of research would limit it to activities done at sea, thus including such aspects as sea trade and travel, maritime technology, and geographic constraints. This however is clearly unsatisfactory, for no one lives permanently at sea and uninfluenced by land-based society. Chapter Six in this book discusses Hindu attitudes to seafaring; these were of course formulated by land-lubbers and were inextricably connected with "internal" Hindu society. Thus a discussion of this apparently important influence on Hindu travel by sea has to consider medieval Hindu norms and prohibitions. Similarly, does maritime history cease to take cognizance of a ship, its cargo and crew once it ties up in some port? Clearly it cannot. Rather, the production of timber for the ship's construction, a land activity, is of relevance. The recruitment of the crew from their (terrestrial) homes must be considered. On a hajj ship we need to look at the origin and motivations of its pilgrim passengers. How can we avoid discussing the production of the goods traded on the ships? For example, the great famine of 1630-32 in India, clearly a land event, crippled sea trade from Gujarat for some years. Equally obviously, land politics sometimes affected maritime matters : Surat's merchants at times refused to sell goods to the English, Mughal nobles occasionally hindered the trade of the Dutch; both events have clear maritime repercussions.

Some of these matters are far into "internal" Indian history. Mughal court politics could at times affect sea trade, but can the maritime historian be expected to do original research on factions at court? Where in fact does one draw the line? No doubt this is a common historiographical problem (if one's subject is the Indian national movement from 1935, how far back does one go to provide essential background?) yet it is a particularly acute one for those seeking to promote and justify a fairly new field of historical study. The only solution, and no doubt an unsatisfactory one, is that there really is no alternative to a thorough evaluation of land influences on sea affairs. The maritime historian has to go as far inland as is necessary, in a functional sense, for adequate evaluation of nautical matters. Thus court politics which, either directly or indirectly, influence sea trade have to be discussed. But need we consider the Safavid-Mughal competition over the strategic fortress of Qandahar? Ostensibly this is a purely "land" matter, of no interest to sea-farers, yet on reflection this also cannot be ignored. This rivalry, and the resultant military operations, hindered the overland horse trade from the Persian Gulf area to north India. Horses being essential for the military elite, they had to come by an alternative route, in other words by sea.

Indian maritime history must, therefore, be seen as a branch of Indian history, and incapable of an autonomous or isolated existence. A truism no doubt, yet this raises, apart from the definitional and limitation conundrum, a mechanical problem of access to research materials. A maritime historian, to do his research, will read documents in one or more European languages but for the medieval period of Indian history various languages, and especially Persian, are essential. The completely equipped maritime historian thus needs to control English, Portuguese, Dutch and Persian at a minimum, and ideally also several other European and Indian languages : Fr. Schurhammer's important researches into the life of St. Francis Xavier are in German, as are many of Fr. J. Wick's important contributions; the fascinating autobiography of the north Indian Jain merchant Banarasidas is in a version of early Hindi called "the dialect of Madhyades"; Zain-ud-din's chronicle *Tohfut-ul Majahideen*, essential for sixteenth-century Malabar, was written in Arabic, and the best translation is into Portuguese. It is true that some translations exist, but with many of these there are serious doubts as to their accuracy and completeness. As just one example, the Hakluyt Society edition of the important account by J.H. Van Linschoten is based on the English translation of 1598, which was quite unsatisfactory and had many omissions. In any case, vast bodies of documentation are not, and will not, be translated.

No doubt every scholar is faced with some linguistic deficiency, and there are also the problems of having the time to do original research on every necessary aspect of any particular theme, and of time and financial limitations on anyone's ability to use all available repositories of material. The obvious solution, which fits well with my view of maritime history as part of Indian history, is cooperation. By this I do not mean just using other people's published work, but rather actual joint scholarly efforts, between people with different language skills, and different theoretical backgrounds. Take, for example, the port city of Cambay from the fourteenth to the seventeenth centuries. The following skills and training would be needed to do a complete study of this topic : an historian, an urban specialist, an historical archeologist, and a geographer, controlling the following languages : Arabic, Persian, Gujarati, Dutch, Portuguese and English. This sort of cooperation seems to me to be not an ideal prescription, but rather a prime necessity.

Where does this leave maritime history? It seems to me that we have much to offer, and much to gain. We can contribute an awareness of nautical matters, for example an evaluation of the conduct and dimensions of sea trade, of India's medieval international contacts by sea, of the lives of Indians overseas, of the pilgrim traffic to the Red Sea, of the real impact of the early sea-based Europeans on Indian history. These, and many others, are not negligible subjects. Yet we must not make too grandiose claims. My own research, especially in *Merchants and Rulers in Gujarat* (Berkeley and New Delhi, 1976), convinces me (Admiral Mahan and K.M. Panikkar to the contrary) that the influence of maritime matters on the land, really throughout the whole course of Indian history, has been less than the other way around. Indian civilization, whether Hindu, Buddhist or Muslim, has drawn remarkably little from maritime influences. No Indian state of any size has derived very significant revenues from seafaring activity. Naval power, even dominance, has had very little influence on land states. Thus the Portuguese in the sixteenth century did control the seas around Gujarat, yet this had precious little impact on the history of Gujarat. One or two nobles whose base was a port city—Malik Ayaz in Diu, Khwaja Safar in Surat—were detrimentally affected by the Portuguese, and on occasion a noble's pride was hurt by having to take a passport, but the total political impact was minor even in Gujarat, let alone in the rest of India. Nor apparently were there profound economic consequences from the Gujarati merchants' acceptance of Portuguese control of their trade. They simply paid the extra duties demanded, sailed more or less under Portuguese auspices, and presumably raised their prices a little to absorb these extra costs. Obviously maritime history can make important contributions, but we should not, and need not, overstate our case.

In the holistic view to which I subscribe, the emphasis will always be on contributions to a whole, not on discrete studies. Thus we can raise problems, based on our maritime research, whose solution is to be found in "internal" Indian history. Sometimes, it is to be hoped, we can introduce dilemmas whose solution will require creative new research in "internal" Indian history. Thus I argued in *Merchants and Rulers in Gujarat* that an explanation of the Gujarati response to the Portuguese can only be found within the internal dynamics of medieval Gujarati society and politics. Similarly, Chapter Six in the present collection raises a problem, basically that of observance of ritual prescriptions. The solution is to be found in the nature of medieval Hinduism, and it is relevant to the discussion of language competencies above to note that this solution will come from a scholar who has a strong background in Hindu legal and religious history, and who reads Sanskrit.

Defining Indo-Portuguese history raises different, but related, problems. The first and most obvious area in which we need clarity is to decide whether this term refers to the history of Goa, or to all areas ruled by the Portuguese, or to all areas in which the Portuguese were influential. Posed in this way the question is perhaps overly schematized : it is rather a matter of a continuum. Nevertheless, and partly as a result of the sorts of records we have available, there are differences. Thus Goan history can include detailed studies of land revenue systems, as B.H. Baden-Powell, and more recently T.R. de Souza, have shown. (See the bibliography at the end of Chapter One for references to these and other studies). It is relatively easy to define Goan history, for the Portuguese ruled a specified area, on which they had a pronounced impact. There were as a result clear differences between life in Goa, especially in the Old Conquests, and life in the neighbouring areas of Konkan. The same applies, though to a lesser extent, in other Portuguese-ruled territories. Yet neither of these two categories exist entirely independent of surrounding Indian areas, just as the Portuguese population was not completely immune to influences from the Indians among whom they lived : thus Chapter Three does not try to discuss the Portuguese crowd in total isolation from the Indian environment in which they operated. This tendency is compounded as we look at the third category of Indo-Portuguese history in areas not ruled by the Portuguese. Here we are back to the problem of where to stop, already discussed in relation to maritime history. Clearly the Jesuits at the Mughal court cannot be seen in isolation from the context of religious discussions at court, of the religious attitudes of the various emperors, or indeed without considering the basic attitudes of Islam to other religions. Nor can we ignore factions at court which at times used the *farangi* priests for their own internal, and mostly political, purposes. Similarly, the activities of private Portuguese in the Bay of Bengal area have to be seen in the context of political and economic changes occurring in this area. Here again the only solution is cooperation with other scholars.

More important is the basic nature of the whole enterprise. It is significant that the seminar is on "Indo-Portuguese" history, not Portuguese colonial history. This change is in part no doubt a reflection of political changes in Goa and later Portugal itself, but more generally it derives, I think, from the general world wide decline in interest in colonial history. The Dutch scholar H.L. Wesseling has considered this problem in his article "The Leyden Centre for the History of European Expansion : Balance and Perspectives", *Archipel*, No. 17, 1979. The rhetoric of politicians opposing colonialism or neo-colonialism has percolated up or down into academia and led to many fewer studies of colonial powers and colonists in isolation from their Asian and African spheres of operation.

In terms of the relations between colonial, and Indian, history there seem to me to be three analytical categories. First there is colonial history per se, a study of Europeans in Asia which concentrates exclusively on the Europeans. Leaving aside present-day political influences, this sort of study is intellectually unsatisfactory because it is illogical to deny any influence at all from the Asian environment on the Europeans living there. No one is entirely an island; every European, even a Portuguese or English viceroy, has to accommodate to some degree with the setting in which he is placed. (This was a major theme in Blair B. Kling and M.N. Pearson, eds., *The Age of Partnership : Europeans in Asia before Dominion*, Honolulu, 1979). To ignore this seems to be extreme ethnocentrism. Fortunately such flawed history is seldom met with today.

It is instructive in this context to consider a recent large study of the Portuguese empire, Bailey W. Diffie and George D. Winius, *Foundations of the Portuguese Empire, 1415-1580* (Minneapolis, 1977). The section dealing with the sixteenth century is by Winius, and he writes in his preface "I have written a history of Portuguese power in Asia, not a history of Asia at the time of Portuguese expansion, or even a full-scale history of Portuguese Asia". Obviously one should write the sort of history one feels is worthwhile, and Winius is to be congratulated on making clear his intentions. Yet my comments in the previous paragraph are unfortunately fully borne out in his work, for it is almost completely one-dimensional. The Portuguese are seen in isolation from the Asian context in which they operated; indeed, when the context is considered at all Winius often gets things wrong. Thus we hear of "Bijapurese", of the "sudra, or commercial caste", of the period between "May and September—the season when the monsoons allowed coastal sailing". Achin and Atjeh are apparently different places, Socotra and Sofala are confused, and Chaul is located in the Gulf of Cambay.

In part these errors result from his relying almost exclusively on Portuguese records; his bibliography shows virtually no acquaintance with either primary or secondary Indian work. What we have is an up-to-date politico-military chronicle vitiated by a failure to present the Portuguese in the round, in the setting in which their deeds took place. A comparison with Holden Furber's volume in the same series, *Rival Empires of Trade in the Orient, 1600-1800* (Minneapolis, 1976), or with C.R. Boxer's *The Portuguese Seaborne Empire, 1415-1825* (London, 1969) is unfortunately very much to the disadvantage of Winius.

The second category grew out of dissatisfaction with the first, and is concerned to describe the impact of the colonial powers on Asian society, and conversely the influence of the Asian setting on European politics and people in Asia. There are of course many degrees within this. Thus some scholars, while not denying the above, prefer still to work primarily from the angle of the colonial power : many distinguished works have resulted, most obviously those of C.R. Boxer, and more recently A.R. Disney's *Twilight of the Pepper Empire* (Harvard, 1978).

The nature and extent of the European impact is of course the prime question, and it requires careful evaluation in an atmosphere as free as possible from current political debates. We need careful delineation of exactly when, and in what areas, there was an impact, and not simply assumptions that the mere fact of the arrival of the Europeans, in and of itself, "made a difference". There is no "Vasco da Gama epoch of Asian history", yet in some areas at sometimes the Europeans obviously did influence some parts of Asian life. To deny this would be as false as to deny the reverse, that Asian people and societies influenced Europeans. Nor is it any contribution to react to nineteenth-century European denigrations of Asia by simply reversing the coin and seeing the Europeans as mere barbarians and the Asians as heroic freedom fighters. This may not be any more accurate than the reverse, and more important it may still give the Europeans greater significance than they really had. This theme underlies Chapter Two in this collection.

We need in fact to be very sensitive in our handling of the nature of colonial impacts. H.L. Wesseling has edited a whole book on the problem, called *Expansion and Reaction : Essays on European Expansion and Reactions in Asia and Africa* (Leiden, 1978). The very word may be unfortunate and head us in the wrong direction, for "impact" implies there is an active force (European colonialism) crashing into a static entity (Asian society), and this clearly does a grave injustice to various dynamic internal developments in Asia. And of course there is also the reverse, a study of Asian influences on Europe, which have been so comprehensively studied by D.F. Lach.

There is a third analytical category, equally deserving of attention. This is to attempt to go one step further into internal Asian history, and to use the European records for what they tell us about actual Indian, or Asian, history. This sort of study is attempted in Chapter Six of the present collection, and these three categories really underly the structure of *Merchants and Rulers in Gujarat*. In that book Chapter Two is purely colonial, dealing with the Portuguese system in Asian waters. Chapters Three and Four deal with the effect of this system on various levels of society in Gujarat, while Chapter Five moves right into Gujarat and tries to explain the response to the Portuguese in terms of Gujarati political and social structures. The stance or angle of vision thus swings, perhaps awkwardly, 180 degrees, from total Portuguese-centrism to total Indian-centrism.

I would stress that while I consider the first category to be unacceptable, the second clearly is valid : it is not necessary to take the third step to write really valuable history. Indeed, of the essays in this book, only Chapter Six really goes to the third stage, and then only tentatively. The others, Chapters Two, Three, Four and Five, simply attempt to look at the Portuguese in an Indian context, to see, for example, their crowds influenced by the Indian environment, their economy in many ways dependent on Indian cooperation.

In sum, I hope that Indo-Portuguese history can contribute, either by implication or more directly, to the wider field of medieval Indian, and Portuguese, history. It should be a symbiotic relationship; I am convinced that we have something to offer, but we also have much to learn. If, for example, we are writing about the impact of the Portuguese on Goan society we must know something of pre-Portuguese Goan society, and of Hindu society in the rest of the Konkan area. If we are interested in conversions we must first study not only the history of the Jesuits but also the nature of medieval Hindu society. We must be prepared to borrow techniques and methods from other historians of India, and generally from other social sciences. The many studies of Indian nationalism could inform an analysis of the Goa Freedom Struggle; work on the important role of Indian subordinates in the British Indian administration has obvious implications for studies of the administration of Goa; recent over-time village studies by both historians and social anthropologists could be duplicated in Goa—indeed surpassed, given the wealth of data from very long time spans available. K.N. Chaudhuri's *The Trading World of Asia and the English East India Company, 1660-1760* (Cambridge, 1978) should be, for its methodology at least, required reading for all who would work on Goan based trade; recent anthropological and sociological work on culture contacts and abrasions are obviously directly relevant.

Some consideration must also be given to the nature of the Portuguese records on which most of this book is based. The subject is considered more fully in Chapter One, but it may be noted here that the problems associated with their use are closely related to the three analytical categories sketched above. Clearly it is easiest to use them to write about the activities of the Portuguese in Asia, for this is what the chroniclers, and the letters of the viceroys, our two major sources, were interested in. Yet even in this area Portuguese records present problems. Most of what we have was written by the elite or for the elite. Thus the activities of *soldados* or *casados* are ignored, or belittled, while *fidalgos* figure prominently. This disparagement was a particular problem for me when I tried to write about the crowd in Portuguese India in Chapter Three, and perhaps I reacted too strongly to the jibes of the elite and became too sympathetic to the lower classes. Certainly the editors of the book in which this piece first appeared thought so; they noted that I offered "a remarkably sympathetic defense of those responsible for the near-chronic unrest in those distant Portuguese outposts".

Apart from this slant, there is the problem that many records which would presumably have dealt more fully with the lower Portuguese orders have not survived. Reports from the captains of the various forts of Goa, or accounts of private Portuguese living outside Portuguese-ruled areas, would be invaluable : for the most part all we have are summaries of such reports, filtered through the perceptions and biases of the elite, in the viceroy's letters to the king.

The problems associated with Portuguese records multiply when we come to the second category and attempt to place them in their Asian context. The Portuguese, or at least the elite whose records we must by and large rely on, tended to despise Indians and other Asians, and either ignore them or criticize them in their documents. Thus to most Portuguese Hinduism was a barbaric superstition worthy of condemnation but not of description. Several chapters in this book, and especially Chapter Five, describe Portuguese dependence on Hindu economic power, and Dr. T.R. de Souza and Prof. G.V. Scammell have also written on this subject. However, much of this, such as Portuguese employing Indian agents, was done in the face of official discouragement or even prohibition. The consequence is, of course, that the records either try to ignore this Hindu role in the Portuguese empire, or mention it only to condemn it. Chapters One and Two investigate in more detail this sort of difficulty. In sum, the ethnocentrism of most of the surviving documentation makes it difficult to use it to describe the undoubted accommodations and compromises the Portuguese had to make when faced with Asian realities.

This serious problem is of course compounded when we try to use Portuguese records for Indian history. They tended to despise the Indians with whom they came into contact—the common derogatory term is "pretos", best translated as "niggers"—and by and large they had no interest at all in edscribing Indians with whom they had no direct contact.

Much fuller discussions of this sort of perceptual problem are to be found in Donald F. Lach, *Asia in the Making of Europe*, 2 vols to date, (Chicago, 1965—) and very urbanely in G.V. Scammell, "The New Worlds and Europe in the Sixteenth Century", *The Historical Journal*, XII, 3 (1969). Here we need only note that Malabaris are denounced because they were enemies, but no systematic attempt is made to describe their society or religion. The Mughals are of interest when they threatén, or when they may be susceptible to conversion, but not otherwise. Dramatic changes occurred in India in the sixteenth century—the Mughal conquests, the spread of bhakti influences, great artistic accomplishments—but unless these developments affected the Portuguese directly they were ignored.

To be sure, part of the problem here is simply the nature of the Portuguese presence in sixteenth-century India. One cannot, after all, describe something one knows nothing about. The Portuguese presence was almost exclusively a coastal one, so that their records deal only with sea-faring and littoral societies. In the next century the Dutch and the English, at this time trading on a basis of equality in India rather than trying to copy the Portuguese and attempt to enforce a maritime monopoly, perforce had to send agents up country, and investigate economic, and some social and political, conditions in the production areas. Thus we have the *English Factory Records*, so much more wide ranging in their concerns than comparable Portuguese documentation. Similarly, there is no Portuguese equivalent of F. Pelsaert's extended account published in English as *Jahangir's India*.

In the sixteenth century, the people who did venture inland and did leave accounts of Indian society (excluding the renegades, who worked for Indian powers but have left no records) were the missionaries. Many of them, notably the Jesuits, were acute enough to work from the maxim of "know your enemy", and thus they tried to understand Indian society and Hinduism the better to make converts. Their accounts are, by definition, biased and hostile, but they are still valuable, and thanks to the labours of many scholars, especially today Frs. Wicki and Silva Rego, they are gradually being made accessible.

The corollary of all this is that Portuguese and missionary

records alone are not enough for those working in the area of Indian history. Clearly such indigenous accounts as may be available must be used. Indeed, scholarly concerns aside, it would be ethnocentric not to do so. And so we come back to the need for collaboration, or at least cooperation. Even the missionary records are in many languages. Fr. Wicki usually writes in German or Latin, and the *Documenta Indica* series he is editing contains documents in all the Romance European languages, including Latin.

It must be stressed again that we as Indo-Portuguese historians should see our relationship with Indian, and secondarily with medieval European history as symbiotic. This means that we must be aware of research findings from Indian specialists, and from scholars working on European society. The Portuguese who came to India were clearly not empty vessels : their minds, attitudes, responses were created in the Europe of the Renaissance and Reformation. Clearly therefore we need to take account of relevant empirical data from medieval Portuguese and Indian history. But we need to go further than this. Profound insights, as noted above, can be derived from various tangentially related work in Indian studies. Similarly, the brilliant social history written by the *Annales* school has much to stimulate and inform our efforts, especially concerning influences on the world views of the Portuguese.

This sort of academic cross-fertilization is clearly apparent in several of the pieces in this collection. Chapter Three draws explicitly on George Rude's work; the whole theme of cooperation and collaboration between European colonialists and indigenous groups, in Chapter Five, has a long history, not just as regards India but also in most other areas subjected to European imperialism; much of the terminology used in Chapter Four draws on the results of discussions at a conference on the Asian port city in California in 1976. In several of these chapters I try to distinguish between theory and practice, between the orders from Lisbon or Madrid and the actions of viceroys and captains in India; again this theme has a long pedigree in many different areas of historical research. We must be prepared to be magpies, and pick up useful information and ideas from many areas and disciplines.

With these sorts of considerations in mind, there are a host of future areas of research available, using mostly Portuguese records. In Chapter One I discuss some topics which could be researched primarily from the Goa archives. More generally, and using again our three analytical categories, the following themes come immediately to mind. In the area of the Portuguese themselves, we need careful studies of the elite, of their "mind-sets", based on their social and political backgrounds in Portugal. What sorts of criteria determined the selection of a viceroy? What mental baggage did he bring with him to India? What sorts of books had he read? Were the Portuguese elite in India from rural or urban backgrounds, had they been involved in Africa, had they travelled outside Portugal before, were they primarily courtiers, what contacts had they had already with non-Europeans? The existence of tensions between religious and secular authorities in Portuguese India is generally recognized, but has yet to be thoroughly worked out; indeed, a good start would be a close investigation of the validity of this perceived dichotomy in the first place, for sixteenth-century Portuguese did not necessarily see things the way secular-minded twentieth-century historians do. In the area of official Portuguese policy, we all accept as a truism the mixing of economic and religious motives encapsulated in the "Christians and Spices" reply, but again this has not yet been systematically worked out over the sixteenth century. At what times, and why, was there a conjunction of these two aims, and when were they incompatible? In which situations did God triumph over Mammon, and vice versa?

So far we know virtually nothing of lower class Portuguese in India. Who really were the casados? Who did they marry, how many children did they have; what occupations did they take up, how did they relate to their Indian neighbours?—all basic questions, all so far unanswered. And what of those even more shadowy figures, Portuguese living outside Portuguese-ruled areas? What can we find out about the several hundred Portuguese living in Gujarat, notably the port of Cambay? What about all the other Portuguese who "went native" in most of the littoral areas of the Indian Ocean? Who were the renegades, the invaluable military experts who served so many sixteenth and seventeenth-century Indian rulers? A.R. Disney quotes a figure of 5,000 of them only between Bengal and

Macassar in 1627. Clearly the total in all Asia is much more. What made them desert, how were they treated in India; indeed in a prenational era is "desert" even the correct word to use? No doubt some of these topics will be impossible to research for lack of data, but for many others it is simply a matter of asking intelligent questions and then trying to answer them.

In the second analytical category, which clearly overlaps with the first, there is still much work to be done in the area of Indian cooperation, despite T.R. de Souza's efforts and my own venture in this field (Chapter Five). G.V. Scammell has contributed an interesting comparative dimension, but there are many areas and periods yet to be explored. Two other major areas of research await sophisticated modern treatment : conversions, and race relations. For both the data is more than adequate, yet so far we really only have descriptions of the former. No one has tried yet to use insights from anthropology (surely the debate over sanskritization is relevant?) from sociology or from psychology to explain why people convert. With few exceptions, shrill statements of previously formulated bias have driven out objective research in this delicate area. As to race relations, again propaganda has replaced research, helped by official backing for one point of view during the Salazar-Caetano period.

In fact there is, it seems to me, a really fascinating problem here. The Portuguese records themselves make clear the depth and extent of Portuguese racism; indeed I think one has to accept that colonialism per se is by definition a racist activity, let alone such clearly ethnocentric elements within it as the *cartaz* system and its justification. On the other hand we do have the survival, often noted by C.R. Boxer, of elements of Portuguese society (language, religion, food, etc.) long after their political power had vanished in areas like Sri Lanka and parts of India and southeast Asia. Are these two phenomena compatible? Certainly we should look at stratification in Portuguese India, both among the Portuguese and between Portuguese and Indians, in terms not just of colour and religion but also in terms of class. Nor should we necessarily seek strict categories : there is perhaps a continuum going from

Portuguese born in Portugal to non-converted Indians (though even within these two extremes there are pronounced class and status distinctions). Does the Catholicism of the Portuguese, as compared with the Protestantism of the Dutch and English, have something to do with race relations? Again, official Portuguese policy encouraged conversions, while Dutch and English policy was if anything opposed : does this affect race relations? Perhaps we should try to see the Portuguese as premodern themselves (and here a deep knowledge of sixteenth century Portuguese society in Portugal is necessary), as "closer to the ground" than the later Europeans. These are some of the questions : the data is available to answer them.

As to the third category, and again there is obvious overlap, the missionary records have yet to be fully used for what they say about Indian society. Nor have the *Decadas* yet been adequately correlated with indigenous sources to illuminate the political history of much of sixteenth-century western India, and especially Gujarat. Diplomatic practice and conventions in medieval Asia are of increasing interest (Prof. M. Athar Ali of Aligarh Muslim University is writing a book on this at present); the Goa records contain valuable information on this topic, much of it published in P.S.S. Pissurlencar's *Agentes*. Finally, we do have available land revenue records for parts of Goa, notably some villages in the Old Conquests, from the late sixteenth century. These are a unique data base in all Indian history, and the results of their analysis would be invaluable not just for Goan history but for Indian agrarian and social history generally.

This sort of listing could go on almost indefinitely. I have, for example, said nothing of possible studies of religious specialists, or of the urban history of Portuguese India. The field is inviting, and still largely untouched. What is needed are linguistically competent scholars asking the right questions, with some comparative skills, and with some awareness of the theoretical implications of Indo-Portuguese history. Then the recent most encouraging beginnings will be built on and work of value in several disciplines and fields will appear.

Bibliographical Note

I have listed in the acknowledgements the places in which these pieces originally appeared. However, I have made numerous changes to these original versions, partly to up-date them, and partly in order to avoid duplication and to achieve internal consistency. There is no doubt still a little duplication in this book : sometimes the same data has been needed to support more than one argument. The footnotes have been standardized, and keyed to the list of abbreviations which follows this note. References to the Portuguese chronicles are, as usual, to decada, book and chapter number; all references to documents in the Historical Archives of Goa are keyed to Pissurlencar's *Roteiro*. All monetary units are expressed in rupees : I have taken a rupee to be 200 reis, a xerafim 300 reis, and a pardao 360 reis. I also occasionally use the Indian terminology of a lakh (100,000) and a crore (10,000,000). These conversions to other currencies are crude enough no doubt, but something had to be done in the interests of comparability, and in any case many of the numbers we do have are frankly very rough indeed, so that rather crude translations to other currencies really only further distort already sometimes dubious statistics.

In the chapters which follow, few changes have been made to Chapters Two and Three, though the title of the latter has been shortened. Chapter Five represents an amalgamation and rewriting of two articles : "Indigenous Dominance in a Colonial Economy", and "Wealth and Power". The piece which now appears as Chapter One has been substantially expanded and updated.

Of the remainder, the Preface is of course new, but some of the ideas in it were first presented to the "Seminar on Problems of Goan Historiography" at the Xavier Centre of Historical Research in Goa in February 1980. A preliminary version of Chapter Six was read at the International Conference on Indian Ocean Studies at Perth in August 1979, and printed in its Proceedings. Comments at that time, and subsequently during several presentations of it in India, led to very major changes and the present version. Similarly, an early version of Chapter Four was read, in absentia, at the conference "The Colonial Port City in Asia" held at the university of California,

Santa Cruz, in June 1976. A two-page summary appeared on pp. 8-10 of Dilip Basu, ed., *The Rise and Growth of the Colonial Port City of Asia* (Santa Cruz, 1979). A greatly modified version of the conference paper was presented at the Goa conference cited above, but the present version, drawing especially on comments from the 1976 Santa Cruz seminar participants, represents a further revision.

ACKNOWLEDGEMENTS

I AM grateful, first of all, to Dr. Teotonio de Souza, Director of the Xavier Centre of Historical Research (Goa), for accepting this book for publication in the *XCHR Studies Series*. Most of the final preparation of the manuscript was done while I was a Visiting Fellow at the Australian National University, from March to July 1980. For generously providing facilities, I must thank Professor D.A. Low, the Vice-Chancellor, Professor Wang Gungwu, Director of the Research School of Pacific Studies, and Mrs. Margaret Carron of the South Asian History Section. Many thanks to the typists—Maria Giuffre, Bernice Ryan, and Margaret Hall—for fast efficient work. As noted in the Introduction, parts of this book are based on previously published articles, a list of which follows. I am grateful to the original publishers for their permission to reprint, usually in greatly revised form, these pieces.

My thanks to : The University Press of Hawaii, for "Corruption and Corsairs in Sixteenth-Century Western India : A Functional Analysis", from *The Age of Partnership : Europeans in Asia before Dominion*, ed. Blair B. Kling and M.N. Pearson (Honolulu, University Press of Hawaii, 1979), pp. 15-42.

Librarie Droz SA, for "Indigenous Dominance in a Colonial Economy : the Goa Rendas, 1600-1670", from *Mare Luso-Indicum* vol. II (Geneva, Librarie Droz, 1973), pp. 61-73.

South Asian Studies Association of Australia and New Zealand, for "Wealth and Power : Indian Groups in the Portuguese Indian Economy," from *South Asia*, no. 3, 1973, pp. 36-44.

The Editor, Quarterly Review of Historical Studies, for "The Goa Archives and Indian History," from *The Quarterly Review of Historical Studies*, XIII, 4, 1973-74, pp. 205-11.

University Presses of Florida, for "The People and Politics of Portuguese India during the Sixteenth and Early Seventeenth Centuries," from *Essays Concerning the Socio-economic History of Brazil and Portuguese India*, ed. Dauril Alden and Warren Dean (Gainesville, University Presses of Florida, 1977), pp. 1-25.

The publishers of *Portugaliae Monumenta Cartographica* (Lisbon), for drawing the map of western coastal India by Fernão Vaz Dourado (1571) for the end-papers.

CONTENTS

LIST OF ABBREVIATIONS USED IN THE FOOTNOTES

AAU	Arquivo Histórico Ultramarino, Lisbon
ANTT	Arquivo Nacional da Torre do Tombo, Lisbon
ANTTCC	Corpo Chronologico, ANTT
ANTTDR	Documentos Remetídos da India, ANTT
APOCR	*Archivo Português Orien·al*, ed. J.H. da Cunha Rivara, 9 vols, Nova Goa, 1857-77
Assentos	*Assentos do Conselho do Estado*, ed. P.S.S. Pissurlencar, 5 vols, Bastorá, Goa, 1953-57.
Barros	João de Barros, *Asia*, 4 vols, Lisbon, 1945-46.
B. Mus.	British Library, London.
Castanheda	Fernão Lopes de Castanheda, *História do descobrimento e conquista da India pelos Portugueses*, 9 vols, Coimbra, 1924-33.
Correa	Gaspar Correa, *Lendas da India*, 4 vols, Lisbon, 1858-64.
Couto	Diogo do Couto, *Da Asia*, 15 vols, Lisbon, 1778-88.
DI	*Documenta Indica*, ed. J. Wicki, 13 vols. to date, Rome, 1948-
DUP	*Documentação Ultramarina Portuguesa*, 5 vols, Lisbon, 1960-67.
Fazenda	"Assentos do Conselho de Fazenda," in HAG, 10 vols.
Gavetas	*As Gavetas da Torre do Tombo*, 7 vols, Lisbon, 1960-70.
HAG	Historical Archives of Goa, Panaji, Goa.
LM	"Livros das Monções do Reino", HAG.
LMBP	*Documentos Remettidos da India ou Livros das Monções*, ed. R.A. de Bulhão Pato, 5 vols, Lisbon, 1880-1935.

THE GOA ARCHIVES AND INDIAN HISTORY

1

In 1972 I started the original version of this chapter by saying that :

> The Historical Archives of Goa, Daman and Diu are one of the most neglected archives on the subcontinent. So far as I know I am one of only five people affiliated with an American or European university (the other four being C.R. Boxer, J.B. Harrison, A.R. Disney and the late I.A. MacGregor) who have spent any extended period of time in these archives, and few Indian scholars have followed up S.N. Sen's visit to Goa nearly fifty years ago. (The bibliography lists works by all the scholars mentioned in this paper.) The reasons for this lack of interest are difficult to set out, but in the case of Indian historians the main problem has been an inability to read Portuguese, a result of the lack of instruction in this language in the subcontinent. Financial considerations have also militated against Indians spending much time in Goa, as also has the fact that until 1961 Goa was foreign, and often hostile, territory.

It is a pleasure to be able now (in 1980) to note that much has changed; indeed Goan historiography seems to be on the verge of a renaissance. Young Goans are beginning to engage in serious research into the history of their homeland. There are still problems of perspective evident : Goans today tend either to condemn or condone all the actions of the Portuguese in the past. But as Portuguese rule fades into a more distant memory this problem will solve itself. Perhaps even more important, at last some senior Indian historians, in Maharashtra, Gujarat and elsewhere, are encouraging their students to learn Portuguese and do research in Goa. This is a most welcome development, and should contribute exciting new perspectives to the study of the history of the Portuguese in India.

This emerging new generation of researchers will be greatly helped by two new developments in the infrastructure so necessary to support scholars. The archives have always been there, and some support, both individual and for publications, has long been forthcoming from the Calouste Gulbenkian Foundation in Lisbon. What is new is first the organization of a regular International Seminar on Indo-Portuguese History. The first meet in Goa in November 1978, was attended by many leading scholars, and was a great success. The more important papers from this conference have been published by Oxford University Press in Bombay under the title *Indo-Portuguese History: Sources and Problems.* A.R. Kulkarni's address to this seminar appeared earlier in *Indica* (see the bibliography).

The OUP book contains many valuable studies, among them Fr. John Correia-Afonso's account of the Jesuit records in Lisbon's Torre do Tombo, and Luis Filipe Ferreira Reis Thomaz's essay on Portuguese sources for sixteenth century Indian economic history. Apart from these and other discussions of sources, the volume also includes a magisterial survey of writings in this field since 1950 by C.R. Boxer, A.R. Disney's case for a land orientation for our studies, Teotonio R. de Souza's eloquent plea for a history of the common man in Goa, and G. Bouchon's excellent analysis of the uses and problems of Portuguese documentation for the history of sixteenth century India. In this she touches on several of the themes of this present book; essentially she is concerned to use Portuguese and Indian records to write Indian history.

The second International Seminar on Indo-Portuguese History met in Lisbon in October 1980. An ISIPH Newsletter also appears from the Xavier Centre of Historical Research, Goa, edited by Dr. Teotonio R. de Souza. It is earnestly to be hoped that this Seminar will continue to provide avenues for publication, for meeting, and for the exchange of news and information.

The second welcome development is the founding in November 1979 of the Xavier Centre of Historical Research, in Goa. At present operating from temporary premises, the

Centre will soon be established in spacious quarters of its own near Panaji. This will be of immense value as a meeting and research centre for all scholars working in Goa. The XCHR also plans to hold annual seminars on Goan history : the first meet in Panaji in February 1980 was on the theme of "Problems of Goan Historiography". Another activity of the Centre is the publication of research studies : the present work is one such Centre-sponsored work.

Now that Goans and other Indians are working in the Historical Archives of Goa the continuing western neglect of these archives is much less of a cause for concern. This neglect is partly because most non-Indian scholars specializing on South Asian history have thought of Goa as a place for a holiday, not as a research centre. Amongst some scholars who are reacting sharply to older Eurocentric historiography there has perhaps also been a feeling that to use European records, especially those of a recently disliked colonial power, is somehow not quite the same as using "real" Indian materials in "real" Indian languages. What is particularly interesting is the number of scholars who have worked on subjects to do with the Portuguese empire but have been unable to use the Goa archives : Jean Aubin, Vitorino Magalhães-Godinho (whose *Os Descobrimentos e a Economia Mundial*, 2 vols. Lisbon, 1963-71 is the outstanding work in the field since World War II), Anthony D'Costa, Niels Steensgaard, B.G. d'Souza, R.B. Serjeant and G.V. Scammell come to mind at once.

It is necessary first to sketch the size and scope of the holdings in the Historical Archives of Goa. There are at present about sixty thousand volumes of documents, covering the period from 1515 to 1914 and later. An additional body of materials is not yet bound and available for research, and more is being added rather frequently. When centralization is completed the expected total will be one hundred thousand volumes. The sizes of the volumes vary considerably, but on average they are around 300 folios; often only one side of the paper is written on. As to the scope of these holdings, in temporal terms the bulk of them are for the seventeenth, eighteenth and nineteenth centuries. As an indication of the weightage towards especially these last two centuries, it may be pointed out that I have read almost everything for the period 1515 to 1670, but the total number of volumes for this period is only about 160. For the sixteenth century, when the Portuguese Asian empire was at its height, the archives of Portugal are much fuller. In any case, most documents from Goa dealing with the sixteenth century have been published in one or other of the collections listed in the bibliography in this chapter.

In terms of geographical coverage, the Portuguese territory of Goa is obviously treated in most depth, with fewer records for other Portuguese areas such as Daman, Diu, Chaul, Cochin, Coromandel, Ceylon, Malacca, Macao, Africa and the Persian Gulf. For non-Portuguese areas, the Portuguese were more interested in powers contiguous to themselves, in particular the five Deccan sultanates, especially Bijapur, and in the seventeenth century the Mughal Empire and the Marathas. The Portuguese were also vitally interested in sea trade, at least until their sea power was destroyed by the Dutch; there is thus a lot of material dealing with trade and Portuguese attempts to control and tax it. More detail on the sorts of things to be found in the Goa archives can be obtained from works listed in the bibliography. It is, however, appropriate to stress at this point that the Goa archives are in no sense a substitute for other bodies of materials. Anyone interested in the Portuguese in India, from almost any angle, must use not only the Goa archives but also the invaluable Portuguese chronicles, the missionary records, the Lisbon archives, and whatever indigenous records may be available.

Something must be said about the mechanical problems and advantages of working in the Goa archives. Although some local records are in Marathi and Konkani, most documents are in Portuguese so that the first prerequisite is a sound reading knowledge of this language. Oral ability is much less important, for the Portuguese language seems to be fast dying out as Goa becomes more "Indian". For interviewing Konkani is essential, though most of the elite speak English. By comparison with Indian languages, at least for a westerner, Portuguese is easy enough to learn, and there is available a good glossary (by S.R. Dalgado) oriented towards Indian Portuguese.

The pre-nineteenth century documents at Goa are all handwritten, but this does not present a serious problem. Sixteenth century Portuguese was usually written in a script which presents numerous differences from standard modern Roman script; as noted, however, many of the sixteenth century documents are already published, and later documents are usually in clear hands. This advantage is, however, sometimes nullified by the condition of the documents. Until P.S.S. Pissurlencar became archivist in the 1930s the archives were sadly neglected and the documents were abandoned to the ravages of predatory insects and the climate. Despite the careful restoration work done under the direction of Pissurlencar, and later V.T. Gune, over the last forty years, some of the earlier documents are very faded, worm-eaten and stained by water.

There are three other technical problems also. First is the very corrupt spellings of Indian place and personal names in the Portuguese documents. Some examples of the problems this can cause will be found in Chapters Five and Six. Second are the numerous, and at first very obscure, abbreviations used for Portuguese names, units of currencies, and weights and measures. Dr. de Souza's useful article listed in the bibliography is a great help here. Finally there is the problem of Portuguese references to geographic units. As one example, they did distinguish carefully between Gujarat and Cambay as areas, with the former referring to a much wider area. The word "Indian" can have several different meanings.

Pissurlencar's work consisted of organizing, binding, and repairing most of the documents in the archives during his tenure of office. Today they are arranged chronologically, within categories, such as "Customs Houses", "Communities of the Old Conquests", "Extinct Convents", "Decisions of the Financial Council", and so on. Access to these volumes is remarkably simple: supporting letters of accreditation are necessary to gain initial access, but so far the leadening hand of bureaucracy has not weighed heavily on the Goa archives.

The bibliography lists various guides to the archives. The indispensable one is still Pissurlencar's *Roteiro*, but even so this is now out of print, and, given new acquisitions, partly out of date. An urgent desideratum is a complete and up-to-date listing of all the holdings of the archives. The bibliographies in Teotonio R. de Souza, *Medieval Goa*, and in my *Merchants and Rulers in Gujarat* provide general surveys of published and unpublished materials available for sixteenth and seventeenth century Indo-Portuguese history. There are also now available valuable accounts of various sources for Goan history in S.P. Sen ed., *Sources of the History of India*, pages 387 to 551 : these articles are listed separately in the bibliography.

Historians do not live by documents alone, but life in Panaji (formerly Panjim) can safely be recommended. The town is connected to Bombay by air, road, rail and sea, the last being a very pleasant 24-hour voyage down the coast, but which is not operated during the monsoon months. Transport within Goa is on a par with any other provincial Indian town. The area's beaches are justly famed, as are the churches of Old Goa and the temples of Ponda. Western-style accommodation and food are, at a price, readily available, thanks to the area's growing popularity as a holiday resort.

The influx of western holiday makers and so-called "hippies" does create problems for westerners who want to work in Goa. Many foreign visitors and residents indulge in behaviour (especially related to the use of drugs, and nude bathing) which is an affront to middle class Goan (and Indian) values. This has, understandably, produced considerable suspicion, and even antipathy, among the local population. The problem for the researcher is that he or she is automatically lumped in with the rest of the western community, yet good relations with local people are at the least desirable, and often essential for the success of one's work. The only solution is to try and differentiate oneself, especially as regards dress and behaviour, and to live not at one of the beaches but in Panaji itself. This way excellent relations can be established with local people and indeed, as one of the few "straight" westerners in Goa, the researcher can also escape ascription to the academic-diplomatic-air-conditioned caste found in some larger Indian cities.

Two more academic problems must be mentioned here. First, it has to be recognized that the Portuguese, at least those who produced documents, were often racially prejudiced. Indians are usually referred to as "gentiles", "Moors" or "Niggers", and as such were despised. Indeed, even the local Christians were discriminated against. This, however, is not a serious academic problem, for everyone despises someone else; historians presumably are used to allowing for the biases of others (for a fuller discussion see Chapter Two). But this Portuguese attitude did have one serious effect from the scholar's point of view. It dictated that most Portuguese thought it beneath their dignity to enquire into or describe Indian society, religion and culture. Indians, after all, were neither Europeans nor Christians, and so the Portuguese thought that they had nothing to learn from them. While there is extant an enormously detailed description of every Portuguese fort in Asia, with lists of every piece of artillery in every fort, and the salaries and duties of all the officials attached to them, there is nothing remotely comparable for Indian things and Indian people. Thus anyone using Portuguese records for Indian history has first to discount for the obvious racial bias, and second has to proceed more by inference than by using direct, detailed descriptions.

The second academic problem concerns the sorts of records which have survived in Goa. As noted, Goa itself is well served, but the same is not true of other Portuguese areas. For these, almost the only information comes from the letters sent home to the king each year by the viceroy. The viceroy's reports were based on letters he had received during the year from the various forts and trading posts, but these inter-Indian letters have not survived, with few and late exceptions. We also lack, for example, the books of the factors in these coastal areas, and the records of their customs houses. To give two more specific examples of lacunae, both related to my own research interests, no records have survived from Diu until late in the seventeenth century. In the later sixteenth century nearly every Indian ship sailing to or from the Gulf of Cambay called here to get a pass and pay customs duties. The records of this fort would thus be of enormous value in estimating the size of, and fluctuations in, Gujarat's overseas trade, the participants in it, and the products traded. Nor, except for a few passing references in missionary records and Persian chronicles, is anything known of the several hundred Portuguese who lived in Cambay in the late sixteenth century. These men were traders; they married local women and seem to have fitted into the prevailing Gujarati commercial scene very closely. The reminiscences of one of these traders would be extraordinarily interesting, but if these were ever written they no longer survive.

At this point it is appropriate to mention some past researches based on work in the Goa archives, and give some indications of other uses which could be made of them. The bibliography lists several older books on particular topics. These tend to be exclusively descriptive, but they often contain extensive quotations from documents, either in the text or in the ubiquitous "Documental Appendix". The publications of the late archivist, Panduranga Pissurlencar, are widely scattered and often difficult to obtain except in Goa itself. Residents of Calcutta should, however, find many of them in the library of the late Sir Jadunath Sarkar. Only a selection of Pissurlencar's works are listed in the bibliography of this chapter. Most, but not all, of them deal with military and political history, and they all contain voluminous extracts from documents in the archives. In addition, he published several invaluable collections of documents *per se*. More recently C.R. Boxer, the great authority on Portuguese colonial history, has worked several times in the Goa archives. His article, "A Glimpse of the Goa Archives", is an excellent introduction to most of the more important runs of documents. His book, *Portuguese Society in the Tropics*, is concerned with the municipal councils of four Portuguese colonies, including Goa, while his very recent *Portuguese India in the Mid-seventeenth Century* provides a remarkable overview of the three themes of conquest, navigation, and commerce. These works are, of course, primarily in the category of Portuguese colonial, rather than Indian, history. The same applies to A.R. Disney's recent, and excellent, monograph *Twilight of the Pepper Empire*, which uses some Goan materials. The "new wave" is best seen in the many articles, and basically important book, of Teotonio R. de Souza.

The potential uses of the Goa archives are numerous. A basic distinction again needs to be made between their value for interior, Goan, history, and their use for topics in other areas of India, and of Asia. They are clearly of prime importance for anyone interested in the Portuguese imperial effort in Asia, and especially in the lugubrious Lusitanian decline in the seventeenth and eighteenth centuries. The imperial aspect of the most important period of this decline has been excellently covered by A.R. Disney, and, from a different perspective and using mostly published Portuguese sources, by Niels Steensgaard. For Indian history, there is some political and military information on the five Deccan sultanates, and the Mughal expansion southwards in the seventeenth century. Many documents concerning these topics have already been published, but they have yet to be analyzed or synthesized with Indian accounts. The Portuguese were reasonably well-informed about events in the Deccan, mainly because they maintained spies at the important Indian courts in this area. More overtly, they sent frequent embassies to these courts.

Much more important, both in terms of quality and quantity, are the Portuguese records concerned with Maratha history. The best guide to these records is V.T. Gune's article in S.P. Sen ed., *Sources of the History of India.*

The Portuguese had intimate relations with Shivaji and his successors; indeed Goa was attacked by them, and the Marathas conquered the important Portuguese area of Bassein. Here again many of the relevant documents have been published by Dr. Pissurlencar, but again the information in them has yet to be used except for the construction of a narrative. It must, however, be kept in mind that, for reasons already outlined, most of those documents on Deccani history are concerned exclusively with diplomatic and military history.

For extra-Goan economic history, the archives are useful primarily for sea trade. Here there are some very detailed accounts, such as the complete cargoes of two Gujarati ships captured in 1630 after they had loaded at Hurmuz. The cargoes and ships were sold at auction in Goa; notable in their contents were horses, rose water, silks, a red dye and a vast quantity of dates, but also sold were three "moorish daggers" and two bundles of broken glass (sold to the king of Portugal for Rs. 3). Forty carpets, mostly designed for use in palanquins, sold at an average price of about Rs. 105. For this same year we also have a complete list of the goods belonging to individual Portuguese sent home to Portugal on the three ships of that year. These ships carried an enormous variety of goods "splendid and stifling" but mostly the latter. One item stands out as overwhelmingly most important, and this was cotton cloths, nearly all from Gujarat, and totalling nearly four and one half million yards. Other topics in the area of sea trade include the spice trade, especially considerable detail on pepper, and the Portuguese attempt to control and tax Indian sea traders.

The documentation available for Goan history is large and wide-ranging. The difficulty here is the common impression that Goa, especially after the sixteenth century, is atypical, and indeed insignificant, in terms of Indian history. This is a most unfortunate attitude. We should not assume that Goa is unusual and uninteresting just because it was ruled by the Portuguese rather than by the British. The Historical Archives of Goa should be seen as offering a unique data bank of information which could be most provocative for both comparison and contrast with similar, though more limited, materials from British India. Historians of other areas of India are increasingly doing micro-level studies, and Goa is ideally, indeed uniquely, suited to these over very long periods. Tom Kessinger's detailed village-level statistics from the mid-nineteenth century (in his book *Vilayatpur*, Berkeley, 1974) were hailed as unprecedented, but in Goa we have available documentation on the first and last European colony in India. Thus such topics as culture contacts and abrasions, religious interactions and administrative forms, can be surveyed from 1510 to 1961. Economic-communal studies, on the lines of Chapter Five in this book, could also be undertaken. There is a large and almost completely untapped body of land and local government records in the Goa archives—about 1,260 volumes in all, mostly in Marathi and Konkani and covering, with many gaps, a period from 1582 to 1887. T.R. de Souza has begun to use these : the indispensable guide for the beginner is the book by Gajanana Ghantkar. In some cases it would probably be possible to trace the ownership and use of a piece of land from the sixteenth century to the present day. Population statistics for Goa are also comparatively excellent. De Souza surveys these for the seventeenth century in his book, and there are more or less detailed censuses available from the late eighteenth century.

Perhaps most interesting of all would be studies of castes in the Goa area, and here I would consider the Indian Christians to be one, or more accurately several, castes rather than a completely distinct and non-stratified group. To me (see Chapter Five) it seems likely that the Saraswats were economically dominant from the sixteenth century to the present day, but further research may show that this was not the case over the whole period. Anyway, the Saraswats were far from being a homogeneous group. A detailed study of their response to colonial rule over four centuries could be a fascinating topic. Other caste groups are also, of course, to be found in Goa, while the survival of caste among converts to Christianity opens up another interesting area. It should also be noted that this sort of study of local history and society could be researched entirely in Goa, using both the archives and collections in temples. By repute these latter are rich, and certainly they are completely untapped, but the prerequisite here would be sound Portuguese, Konkani and Marathi, and possibly some Persian. The only accounts of Goan temple records with which I am familiar are those by G.H. Khare.

This chapter has perhaps stressed too much the limitations of the materials in the Goa archives. In fact I am convinced despite encouraging developments over the last ten years, that they are still most shamefully underutilized. They will, I hope, increasingly come to be recognized for what they are : prime sources not just for Portuguese areas, but, in conjunction with other materials, for most of south and western India.

BIBLIOGRAPHY

Guides to the Goa Archives

Boxer, C.R. "A Glimpse of the Goa Archives". *Bulletin of the School of Oriental and African Studies*, XIV (June, 1952), pp. 299-324.

Christian, John LeRoy. "Portuguese India and its Historical Records". *Hispanic American Historical Review*, XXV (1945), pp. 140-51.

de Souza, Teotonio R. "Portuguese Records for Indian History at Goa and Lisbon". *The Indian Archives*, XXV (1976), pp. 24-36.

———."Portuguese Source Material in the Goa Archives for the Economic History of Konkan in the Sixteenth and Seventeenth Centuries". S.P. Sen, ed. *Sources of the History of India*, I, Calcutta, 1978, pp. 425-41.

Ghantkar, Gajanana. *An Introduction to Goan Marathi Records in Halakannada Script*. Margão, 1973.

Gune, V.T. *Assentos do Conselho do Estado : A Detailed Subject-Index and a Table of Contents in Brief*, Panaji, 1972.

———.*Guide to the Collection of Records from the Goa Archives*, Panaji, 1973.

———."An Outline of the Administrative Institutions of the Portuguese Territories in India and the Growth of their Central Archives in Goa, 16th to 19th Century A.D." V.D. Rao, ed. *Studies in Indian History : Dr. A.G. Pawar Felicitation Volume*, Bombay, 1968, pp. 47-92.

———."The Portuguese Records Relating to the History of the Marathas", S.P. Sen, ed. *Sources of the History of India*, I, Calcutta, 1978, pp. 387-422.

Humbert, J. "The Goa Archives in 1963". *Indica* I (1964), pp. 47-66.

Meersman, Achilles. "The Goa Archives and the History of the Franciscans in India". *Indica* I (1964), pp. 171-85.

Pissurlencar, P.S.S. *Roteiro dos Arquivos da India Portuguesa*, Bastorá, 1955.

Sen, S.N. *A Preliminary Report on the Historical Records at Goa*, Calcutta, 1925.

Aids

Dalgado, S.R. *Glossário Luso-Asiatico.* 2 Vols, Coímbra, 1919-21.

de Souza, Teotonio R. "A Tentative Check-List of Abbreviations in Portuguese Archival Records". *Indica*, XIV (1977), pp. 117-24.

Taylor, James L. *A Portuguese-English Dictionary*, 2nd ed., Stanford, 1970.

Yule, Henry, and A.C. Burnell. *Hobson-Jobson.* Delhi, 1968.

Published Documents from the Goa Archives—Books

Abranches Garcia, Jose Ignacio de, ed. *Archivo da Relação de Goa, contendo varios documentos dos seculos 17, 18 e 19.* 2 Vols, Nova Goa, 1872-74.

Bragança Pereira, A.B. de, ed. *Arquivo Português Oriental.* 11 Vols, Bastorá, 1936-40.

Baião, António, ed. *A Inquisição de Goa. Correspondencia dos inquisidores da India, 1569-1630.* 2 Vols, Coímbra, 1930, Lisbon, 1945.

Cunha Rivara, J.H. da, ed. *Archivo Português Oriental.* 9 Vols, Nova Goa, 1857-76.

Gune, V.T. ed. *Assentos do Conselho do Estado. Supplementary Series*, vol. I, part 2, *Assentos 1624-27.* Panaji, 1972.

Pissurlencar, P.S.S. ed. *Agentes da Diplomacia Portuguesa na India (Hindus, Muçulmanos, Judeus e Parses)*, Bastorá, 1952.

——.*Assentos do Conselho do Estado, 1618-1750.* Bastorá, 1953-57.

——. *Portugueses e Maratas.* 6 vols, Nova Goa and Bastorá, 1926-40.

——.*Regimentos das Fortalezas da India*, Bastorá, 1951.

——.*Portuguese Records on Rustamji Manockji, the Parsi Broker of Surat.* 2 Vols. Nova Goa, 1933.

——.*Tombo de Goa, Bardêz e Salcete por Francisco Paes*, Bastorá, 1952.

Journals containing Documents from the Goa Archives

Boletim da Filmoteca Ultramarina Portuguesa. Lisbon, 1954-

Boletim do governo do Estado da India. Nova Goa, 1857-77.

Boletim do Instituto Vasco da Gama. Panaji, 1926-63.

Bulletin of the Institute Menezes Bragarça. Panaji, 1963-
O Chronista de Tissuary. Periódico Mensal. Nova Goa. 1866-69.
O Gabinete Litterário das Fontainhas. Nova Goa, 1846-48.
Indica. Bombay, 1964-
Instituto Vasco da Gama. Nova Goa, 1872-75.
O Oriente Portuguêz Revista da Commissão Archeologica da India Portugueza. Nova Goa, 1904-20, 1931-41.
Studia. Revista Semestral. Lisbon, 1958-

Some Secondary and General Works Relating to Goa

Aubin, Jean, ed. *Mare Luso-Indicum,* Paris, 1971-
Baden-Powell, B.H. "The Villages of Goa in the Early Sixteenth Century". *Journal of the Royal Asiatic Society* (1900), pp. 261-91.
Boxer, C.R. "Fidalgos Portugueses e Bailadeiras Indianas". *Revista de História* (São Paulo) XLV (1961), pp. 83-105.
——. *Portuguese India in the Mid-seventeenth Century.*
——. *Portuguese Society in the Tropics. The Municipal Councils of Goa, Macao, Bahia and Luanda, 1510-800,* Madison and Milwaukee, 1965.
Braz de Albuquerque, V.A.C. *O Senado de Goa,* Nova Goa, 1909.
Correia-Afonso, John. "Indo-Portuguese History in Jesuit Archives", S.P. Sen, ed. *Sources of the History of India,* I, Calcutta, 1978, pp. 530-40.
——. *Jesuit Letters and Indian History,* 2nd ed. Bombay, 1969.
Cottineau de Kolgneu, Diniz L. *Histŏria de Goa.* Vestido em Portuguèz, e Accrescentado com algumas notas, e rectificações por Miguel d'Abreu. Nova Goa, 1858.
D'Costa, Anthony. *The Christianization of the Goa Islands,* Bombay, 1965.
——. "Tombo de Baçaim, 1610-1730". *Indica* VI (1969), pp. 105-18.
de Souza, Teotonio R. "Glimpses of Hindu Dominance of Goan Economy in the 17th Century". *Indica* XII (1975), pp. 27-35.
——. "Goa-Based Portuguese Seaborne Trade in the Early Seventeenth Century", *Indian Economic and Social History Review* XII (1975), pp. 433-42.

de Souza Teotonio R. "Goan Agrarian Economy in Crisis (1510-1649)". *Itihas*, Journal of the Andhra Pradesh Archives IV (1976), pp. 55-70.

——. *Medieval Goa : A Socio-Economic History*, Delhi, 1979.

——. "Xenddi Tax : A Phase in the History of Luso-Hindu Relations in Goa (1704-1841 A.D.)". P.M. Joshi, ed. *Studies in the Foreign Relations of India*, Hyderabad, 1975, pp. 462-71.

Disney, A.R. *Twilight of the Pepper Empire : Portuguese Trade in Southwest India in the Early Seventeenth Century*, Cambridge, 1978.

D'Souza, B.G. *Goan Society in Transition*, Bombay, 1975.

Ferreira Martins, José F. *História da Misericordia de Goa*. 3 vols. Nova Goa, 1910-14.

Fonseca, José Nicolau da. *An Historical and Archeological Sketch of the City of Goa*, Bombay, 1878.

Gracias, J.B. Amancio. *Subsídios para a história Economico-Financeiro da India Portugueza*, Nova Goa, 1909.

Gune, V.T. "Sources of the History of the Former Portuguese Possessions in India". S.P. Sen, ed. *Sources of the History of India*, I, Calcutta, 1978, pp. 472-529.

Harrison, J.B. "Colonial Development and International Rivalries outside Europe: Asia and Africa". *New Cambridge Modern History*, Cambridge, 1968, III, pp. 532-58.

——. "Europe and Asia". *New Cambridge Modern History*, Cambridge, 1970, IV, pp. 644-71.

Indo-Portuguese History . Sources and Problems, Delhi, 1980.

Khare, G.H. "The Archives of the Vaishnava Matha of Sarasvata Brahmans at Partagali". *Proceedings of the Indian Historical Records Commission*, XXVIII (1951), pp. 50-55.

——. "An Uncared-for Source-Material for the Socio-Economic History of the Goa Territory". S.P. Sen, ed. *Sources of the History of India*, I, Calcutta, 1978, pp. 447-56.

Kulkarni, A.R. "Indo-Portuguese History : Perspectives and Expectations". *Indica* XVI (1979), pp. 221-30.

MacGregor, I.A. "Europe and the East". *New Cambridge Modern History*, Cambridge, 1958, II, pp. 591-614.

Magalhães-Godinho, Vitoríno. *Os Descobrimentos e a Economia Mundial*, 2 vols, Lisbon, 1963-71.

Mathew, K.S. *Portuguese Trade with India in the Sixteenth Century*, New Delhi, 1981.

Moniz, A.F. *História de Damão*, 4 vols, Bastorá, 1904-23.

Pearson, M.N. *Merchants and Rulers in Gujarat : The Response to the Portuguese in the Sixteenth Century*, Berkeley and New Delhi, 1976.

Pereira Nunes, A.R. *Diu. História*, Nova Goa, 1907.

Pinto, Christovam. *As Gauncarias de Goa*, Nova Goa, 1880.

Pissurlencar, Panduranga. *The Attitude of the Portuguese towards Shivaji during the campaigns of Shaista Khan and Jai Singh*, Calcutta, 1927.

——. "The Extinction of the Nizamshahi. (A Survey of the History of the Deccan from Portuguese Sources)". *Sardesai Commemoration Volume*, ed. Shripad R. Tikekar. Bombay, 1938, pp. 27-46.

——. *Goa Pre-Portuguesa atraves dos Escritores Lusitanos dos Séculos XVI e XVII*, Bastorá, 1962.

——. *Prince Akbar and the Portuguese*, Calcutta, 1928.

——. "Tentatives dos Portugueses para a Ocupação do Concão (Séculos XVI, XVII e XVIII)". *Anais da Acadèmia Portuguesa de História II Serie 6* (1955), pp. 425-36.

Priolkar, A.K. *The Goa Inquisition*, Bombay, 1961.

Scammell, G.V. "Indigenous Assistance in the Establishment of Portuguese Power in Asia in the Sixteenth Century". *Modern Asian Studies* XIV (1980), pp. 1-11.

Scholberg, Henry. "The Records of Portuguese India : Some Untapped Sources". S.P. Sen, ed. *Sources of the History of India*, I, Calcutta, 1978, pp. 457-71.

Serjeant, R.B. *The Portuguese off the South Arabian Coast*, Oxford, 1963.

Shastry, B.S. "The Book of Correspondence Concerning Kanara : A Note". *Indica* XV (1978), pp. 121-26.

——. "The First Decade of Portuguese-Vijayanagara Relations". *Journal of Indian History* LII (1974), pp. 147-55.

——. Some Important Published Primary Sources in Portuguese". S.P. Sen, ed. *Sources of the History of India*, I, Calcutta, 1978, pp. 541-51.

——, "The System of Cartazes". *Journal of Indian History* LI (1973), pp. 297-300.

Silva Correia, A.C.G. da. *História da Colonização Portuguesa na India*, 6 vols, Lisbon, 1948-58.

Steensgaard, Niels. *The Asian Trade Revolution of the Seventeenth Century*, Chicago, 1974.

CORRUPTION AND CORSAIRS IN SIXTEENTH-CENTURY WESTERN INDIA 2

A Functional Analysis

ANY HISTORIAN writing about any society at any time is faced with the problem of interpreting the past in its own terms rather than through twentieth century eyes. Western historians writing on Asia have two other difficulties to contend with. First, any Asian society is demonstrably more different from any western society than is, say, English compared with Spanish society. For a westerner writing on Asia the task of getting inside the area is thus apparently much more difficult than if one were to study another western area. Second, until recently western scholars specializing in Asia were all too often fellow citizens of colonial rulers in some part of the area. This colonial legacy, not to mention current neocolonial attitudes, simply aggravates the first problem, which is itself formidable enough.

Perhaps, however, we westerners who write on Asia really have an advantage over our colleagues working on colonial America or Tudor England. For them it is all too treacherously easy to find "nationalism" and other modern phenomena, for they are dealing with societies demonstrably rather similar to those in which they live. But the first thing westerners learn when they start to study Asia is that it is definitely, indeed apparently totally, different. Language, people, geography, culture : everything is strange. Only the most uncouth and slipshod historian can pretend to find "democracy" in ancient India or "feudalism" in the Muslim period (though such discoveries have, regrettably, been made). Thus in theory, though lamentably not always in practice so far, it should be easier to start our studies with a clean slate and a mind uncluttered with extraneous and deceptive cultural baggage.[1]

Historians of South Asia have so far been remarkably un-self-conscious about these problems.[2] They are content to write on, unconcerned to elucidate and analyze the assumptions and prejudices which must colour their work.[3] The purpose of this chapter is not, however, to name names but to suggest an underlying unity in the historiographical problems we face. Every historian is faced with preconceptions stemming from the basic fact that the scholar is situated in the present but writing about the past. Western historians of Asia have in addition to avoid cultural biases. But the basic necessity is to see past societies, whether more or less culturally different, in their own terms and to analyze groups and practices in terms of their functions in such societies. What follows is an attempt at such a functional analysis of a practice, "corruption", and a group, "corsairs", in sixteenth century western India. The treatment of the first in older historical writing shows a failure to see the past in its own terms; the treatment of the second reflects both this and also cultural bias.

Everyone is against corruption today. The word itself is immediately recognized as having derogatory connotations that the official of whom it is used is not fit to hold public office. Late nineteenth-century English writers on the Portuguese empire in India all pointed to the existence of "corruption" in the Portuguese administration as one important reason for the decline of this empire. F.C. Danvers put it like this : "A laxity of Government, and a general corruption amongst the servants of the State, in which each one, regardless of the public interests, sought but his own benefit and the accumulation of wealth, only too certainly prepared the way for the downfall of Portuguese rule in India".[4] R.S. Whiteway, W.W. Hunter, and a little later Vincent A. Smith all agreed.[5] These writers did adduce other reasons for the Portuguese decline : the two favourites were mixed marriages, which produced "a languid population of half-breeds", and "the grip of religious superstition".[6] But a more recent writer, the American F.C. Lane, plumps for corruption as the sole cause of what he regards as Portuguese decline in the later sixteenth century.[7]

Danvers, Whiteway, Smith, and Hunter were all associated with the English Indian Civil Service or its affiliates. All were oriented to what seemed to them to be its higher standards, or at least ideals, as compared to the sixteenth century Portuguese Indian administration. This outlook is made perfectly explicit by Whiteway and Hunter. The former opined that Portugal lacked "that official class which is the backbone of efficient civil administration" and pointedly praised "Albuquerque's business habits—habits that are now, as they were then, the right ones for an administrator".[8] Hunter, in his panegyric to the British empire in India, pointed out that Portugal "represented the reactionary spirit of medievalism, as against the modern methods of the Protestant nations".[9]

These observations obviously refer to the temporal problem of historical writing. The tendency to find the roots of the present in the past—and to condemn the past if it does not live up to the ideals of the present—has, since 1931, often been known as "whiggishness". Herbert Butterfield's classic description of this tendency applies exactly to the work of these four British historians : the tendency to "emphasize certain principles of progress in the past and to produce a story which is the ratification, if not the glorification, of the present".[10]

What were the standards by which a sixteenth century Portuguese administrator in India felt he should regulate his conduct? He was obviously not a member of a modern bureaucracy in the Weberian sense.[11] The difference is seen particularly in the method of selection for office and the whole conception of what holding an office entailed. Appointments to high office were made by the king of Portugal; those to the lower ones were made by his governors or captains in India. They were made on strictly personal grounds, not on the basis of suitability or training. There were two main grounds. First was past meritorious service, usually in a military capacity either by an appointee or his or her forebears. Thus the daughter of a deserving *fidalgo* ("noble"—literally "son of a somebody") would receive from the king as her dowry an office which would go to her successful suitor. Similarly, the great historian Diogo do Couto received his position as official chronicler not so much because he had demonstrated literary ability as because he had served in the Portuguese fleets in India for several years. That he emerged as a productive and skilful chronicler says little for the way in which he was selected.

The second criterion, used increasingly as the finances of the state deteriorated, was purchase : offices were put up for sale to the highest bidder. Here also the basis of selection was personal; in this case the determining factor was wealth rather than past service. Training or skill was not considered, although an informal scheme restricted those who could bid. A non-*fidalgo*, however wealthy, could not bid for the captaincy of a prestigious fort such as Diu.

Not only the selection for office, but also the conception of office, was governed by personal factors. An appointee's loyalties were not to an impersonal conception of an office in which he was to discharge certain functions guided by established rules. Rather, Portuguese officials of all grades treated their offices as a personal service to the king who had appointed them. They owed obedience and loyalty to him personally or to his deputy; so long as the king or governor was satisfied there were no other criteria to be met.

Thus what is at issue here, and what has been obscured by the whiggishness of Lane and the four earlier British writers, is not a "good" administration juxtaposed with a "corrupt" one so much as a modern bureaucracy contrasted with a premodern one. At least in aspiration, the nineteenth century Indian Civil Service conformed to Weber's definition of a "modern officialdom", characterized essentially by the "abstract regularity of the execution of authority, which is a result of the demand for 'equality before the law' in the personal and functional sense".[12] On the other hand, the Portuguese Indian administration was, as noted, far from any ideal of abstract regularity; it was governed, both as regards appointment and performance, mainly by personal factors.

A Portuguese official was appointed to a post as a person, and the post became in effect his property during his tenure. Any distinction between public and private funds was blurred. The bulk of his salary came not in cash from the capital of Goa but in the form of various privileges and perquisites, such as the right to collect a certain tax or to trade on a particular route.[13] These rights were laid down in the official letter of appointment, but they shaded off into other customary rights such as the privilege of being allowed to demand a gratuity before issuing a *cartaz* (a pass for a merchant ship)—which were unspecified but hallowed by tradition. These unofficial but tacitly accepted perquisites in turn merged into outright illegal exactions and abuses, notably demanding forced loans from local traders and trading in goods which were meant to be royal monopolies. The perquisites and the unofficial privileges were regarded by king and governor as unexceptional, but they did try to limit the abuses.

This system of payment was reflected in the prices paid for the various offices. The posts were seen as property from which the holder expected to make a profit, and he thus was willing to pay much more for a lucrative post than for one with few opportunities for pickings, even if the status of the latter was higher than that of the former. In a sale of 1618 more than twice as much was paid for the post of judge of the Goa custom house as for the captaincy of the whole city; the high status attached to the captaincy could not compete with the perquisites, legal and otherwise, available to a customs official. The post of captain of Diu was worth five times and that of Hurmuz fourteen times that of Goa, for these two forts were notoriously profitable.[14]

The prevailing attitude was well expressed by the *fidalgo* who visited a religious house to say good-bye to the inmates before leaving to take up an appointment as captain of a fort. One of the clerics counselled him, "Be content with what is yours, favour the poor, and do justice". The *fidalgo* retorted that he fully intended to get all he could, as did all the others, "because I am not going to my fort for any other reason than to come back rich".[15]

The fuzzy distinction between public and private property can be illustrated by many instances in the sixteenth century. All Indian ships trading out of the Gulf of Cambay in Gujarat were meant to call at the Portuguese fort of Diu to pay customs duties and get a Portuguese pass. Sometimes, however, a Portuguese captain would supplement his salary by taking a bribe from an Indian trader, in return exempting his ship from having to call at Diu.[16] In one particularly flagrant case a Portuguese sea captain used his fleet to protect such a transgressing Indian trader not only from pirates but also from other more law-abiding Portuguese ships.[17] Yet this Portuguese was simply protecting the man who provided a presumably substantial part of his salary.

The most common means of supplementing an income was forced trade; a local merchant would be forced to buy a product he did not want, and at a high price, or sell to the captain or other officials at a low price. Thus one of the official perquisites of the captain of Diu was the right to trade on a monopolistic basis to East Africa. The main item in the return cargo was ivory. This the captain handed over to the head of the local Gujarati merchants, who in turn forced other merchants to buy it.[18] Other merchants were forced to carry goods free on their ships, sometimes to places where they did not want to go. Forced loans were also frequent, sometimes secured by means of arbitrary arrests.[19]

Complaints of these abuses in Diu date from the 1540s, but they do seem to increase in frequency late in the century and into the seventeenth century. A captain of the Portuguese fort of Hurmuz in the 1550s was proud of having made only Rs. 60,000 during his tenure. This was in fact a very substantial sum, but later captains made up to seven times more than this.[20] The apparent general increase is no doubt partly a consequence of better documentation for this period, but it is to be expected that abuses would increase as Portuguese power declined. Diu's revenues, for example, fell sharply in the seventeenth century thanks to Dutch and English competition. The captains of Diu were more and more driven to abuses to make the profits they felt they should gain from their position.

Dubious practices were legion. In the later years of the century captains of forts began taking artillery from the forts to use in their own ships.[21] Throughout the century there are records of Portuguese ships seizing and plundering native ships which had Portuguese passes.[22] An account of the revenues of Goa for 1545 ended by saying : "This is what Goa yielded last year, apart from what was stolen by the officials".[23] A year later an official complained that he had been trying to get opium for the King's trade but had been hindered by the fact that the Portuguese officials in charge of trade seemed more interested in sending this product to Gujarat (with which state Portugal was at war at the time) and other places than in obtaining it for the king.[24] In 1566 another official represented various abuses to the governor and for his pains was forced to flee Goa and seek refuge "among the Hindus and lax Christians of Bardes".[25] Perhaps Diogo do Couto best summed up the state of the administration late in the century when he wrote : "for the king's property to increase it should pass through few hands, and the fewer hands of officials it has contact with the greater will be its increase".[26]

Such instances of what today would be called "corruption" can usually be explained by the confusion in most officials' minds between public and private property. Such confusion was part of every administration at this time : the distinction has been made explicit, and an attempt made to enforce it, only in the last two centuries. Why then the contemporary criticism just quoted? The explanation is that this criticism was of abuses ("abusos", not the modern "corrução") of the system. The governors and chroniclers were not complaining of corruption in the modern sense (for they were not twentieth century men) but of people who went beyond the bounds of custom in taking perquisites. To take the cannon needed for the defense of an important fort was clearly going too far. So was the plunder of a native ship with a Portuguese pass or accepting a bribe for exemption from customs payments. Indeed, according to one contemporary the tyranny of the Portuguese captain of Diu caused a frontal attack on the fort by the neighbouring state of Gujarat : "The captain of the fort caused the seige of Diu because he behaved so badly to the King of Gujarat and the local Muslims that if they had been Christians they would have had good cause to become Muslims".[27] Such abuses were regarded as bad because they could lead to Indian retaliation or increased attempts to evade Portuguese trade control. No one complained because the captaincy of Hurmuz sold for fourteen times that of Goa : the fact of greater perquisites in Hurmuz was part of the system and appears remarkable only today.

Two recent studies seem to be making this same point. A.R. Disney noted "it was of course not unusual in the seventeenth century for viceroys and governors to spend part of their personal fortunes on state affairs, just as it was common practice for them to use the opportunities of office to reimburse themselves". Niels Steensgaard stresses the symbiotic connection between "corruption" and the very functioning of the state, concluding that "A clean-up of the constitutionally determined corruption would have torn asunder the bonds that held the Estado da India together".[28]

There remains the question of the effectiveness of such an administrative system. "Abusos" were clearly dysfunctional, and the governors complained about them, but the whole network of perquisites and influence, of a shadowy line between public and private property, *was* the system : "corruption" was functional in sixteenth century terms. The Portuguese Indian administration was not ideal. It was probably less efficient than a modern bureaucratic structure, or certain other sixteenth century systems, but it is a solecism to criticize the Portuguese administration for its failure to fit an ideal nineteenth and twentieth century model.

If "corruption" in the sixteenth century was less simple than the modern use of the word implies, so also was "piracy" in the southwestern coastal area of Malabar. Natives of Malabar who attempted to trade outside the Portuguese control system, and anyone else in the area who opposed them, were described by the Portuguese as "cossarios" or "Malavares"; the terms were usually interchangeable.[29] A "cossario" (in modern Portuguese "corsario)" is, strictly, a corsair.

Modern Portuguese historians continue to use "corsario" to describe anyone who did not submit to Portuguese control of their trade in Malabar.[30] There is no way of knowing exactly what they mean by this term; presumably something between pirate and privateer is implied. More confusion appears when we consider the doyen of foreign scholars of the Portuguese empire, C. R. Boxer. In a Portuguese work he used "corsario", in an early English work he called them "pirates", and in his standard history of the Portuguese colonial empire he consistently uses "corsair".[31] Again there is some ambiguity, but clearly Boxer does not regard these people as simply pirates.

Two Indian nationalist writers, K. M. Panikkar and O. K. Nambiar, reverse the coin. Nambiar, in his book on the Kunjalis, the admirals of the zamorins of Calicut (the main opponents of the Portuguese in the sixteenth century in Malabar), glorifies these guerrilla warriors and regards them as patriots, even nationalists. The title of the first edition of his work nicely sums up his revisionist aim : *Portuguese Pirates and Indian Seamen.*[32] Panikkar foreshadowed Nambiar's findings and stressed that the Kunjalis were not pirates : they were auxiliaries of the zamorins.[33]

None of these writers attempts to discriminate. One blanket term, whether "corsario", "corsair", "pirate", or "admirals of the Zamorin", is used. All these writers follow the sixteenth century Portuguese accounts in making the Malabaris a monolithic group. All use the same sources; the differing results depend on author's bias. But these Malabaris were undefined only to their contemporary European observers. Grouping them under one (usually derogatory) label is an example of both whiggishness and cultural bias, for in their own, Indian, terms these people were not homogeneous. In functional terms some were pirates, some were corsairs, some were guerrilla warriors and many were inoffensive traders.

What was the sixteenth century context in which these Malabaris operated ? As soon as they got to India the Portuguese tried to monopolize all trade in spices. Throughout the century and later a stream of decrees and instructions from Portugal and Goa insisted that all trade in spices was reserved entirely for the Portuguese crown and its agents. If this monopoly could be enforced the Muslim powers in India, the Red Sea, and Egypt would lose their most profitable trade while the Portuguese could buy cheap in Asia and sell dear in Europe.

The basic policy, in effect a codification of the prohibitions of the previous twenty years, was laid down by the Portuguese king D. Manuel in 1520 and repeated with minor variations throughout the century :

> We forbid and order that no captain, factor, secretary of a factory, or any other official concerned with trade or justice, or any other person of whatever rank or status he

> may be, whether Christian or Muslim or any other, trade, or cause to be traded, or transport, pepper, cloves, ginger, cinnamon, mace, nutmeg, lacre, silk, or borax, nor in any way buy any of these things.[34]

Any Portuguese offending was to lose all his property and salaries; a Muslim was to be imprisoned after forfeiting all his goods and the ship in which the forbidden article was found.[35] Occasionally a closely regulated trade within Asia was permitted, and puppet rulers sometimes were conciliated by being allowed to trade in pepper in limited quantities. Further, the crews of ships returning to Portugal were paid partly by being allowed to take small quantities of spices with them. Nevertheless, Portugal did try to reserve all pepper trade within Asia to agents of the king and to restrict the transport of pepper to Europe to Portuguese ships sailing via the Cape of Good Hope.[36]

This policy was never completely enforced.[37] There was much evasion of the Portuguese system, and in some cases there was open armed opposition to their pretensions. In this opposition the zamorins of Calicut, on the Malabar coast, were prominent, for Calicut had been the great Indian centre for the spice trade in the fifteenth century. If Portugal could enforce its monopoly, Calicut would be ruined while Cochin, formerly subordinate to Calicut and in the sixteenth century a Portuguese puppet, would prosper. The zamorins thus had good economic reasons to resist, and in any case it seems that they simply found Portuguese demands to be unfair. In 1502 Vasco da Gama demanded that the zamorin expel from Calicut all Muslims from Cairo and the Red Sea. The zamorin refused —"for it was unthinkable that he expel 4,000 households of them, who lived in Calicut as natives, not foreigners, and who had contributed great profits to his Kingdom".[38]

Not only the rulers of Calicut, but also her local merchants and traders, resisted the Portuguese claims. The Portuguese wanted to monopolize the trade in Malabar pepper, but this was the main product of the area. The Portuguese forbade trade to Aden and the Red Sea, but this was the main route sailed from Calicut. The Portuguese attacked merchants from these two places; but these were the main purchasers of Malabar's products. The Portuguese forced contracts at fixed low prices on the Malabar rulers for the supply of pepper, but other merchants offered much higher prices. The Portuguese later in the sixteenth century treated all ships from Calicut as pirates and sank them on sight, even ships carrying only provisions, and sometimes ships with Portuguese passes. What choice did the merchants of Calicut, and their ruler, have but to resist ?[39]

Commercially, this resistance took the form of continuing to trade in pepper. The trade direct to the Red Sea from Calicut was blocked more or less effectively by the Portuguese and the great foreign Muslim merchants of Calicut had by 1513 moved out to safer parts, such as the Red Sea, or Hurmuz, Gujarat, and Vijayanagar.[40] There the big traders sat in safety and waited for the pepper to come to them. Reports of direct trade between Calicut and the Red Sea are few, and they disappear completely during the 1530s.[41] From this time the pepper went to the Red Sea from Kanara, north of Malabar, from the Bay of Bengal, and from Acheh in Sumatra.[42]

It was easy enough for the great foreign merchants to move out and let others bring the pepper to them and for the Red Sea traders to get pepper from other production areas. The local merchants of Calicut had no such alternatives, but even so they had little difficulty in acquiring pepper in Malabar. They were helped by Portugal's lack of control over the areas where pepper was grown. In Malabar the Portuguese exercised a tenuous control over the coastal area through their puppet rajas, but none at all inland. Given that the Portuguese tried to pay low prices and paid one-quarter of the price in goods (usually copper) for which the growers had little use, it is not difficult to understand how easily the "illegal" traders found their cargoes. In 1520 a Portuguese official asked an important trader why he had stopped bringing pepper to the coast and selling it to the Portuguese. He replied : "Because I find that there are many people who beg me to sell it, and pay the price I want in my own house, without troubling me to come to the coast."[43]

The Portuguese were dependent for their supplies on the petty rajas of the producing area and the merchants who operated between the growers and the exporters. To conciliate the rajas they paid them regular pensions, but these were not always sufficient to ensure their goodwill and cooperation. In 1549 the raja of the main production area, appropriately called the "king of pepper" by the Portuguese, decided to retaliate against the insults and abuses to which he was subjected by the raja of Cochin. His method was to ally with Portugal's traditional enemy, the zamorin of Calicut. As a result the Portuguese were involved in four years of war in the area, and during this time it was very difficult for them to get pepper to send to Portugal.[44]

The growers of the pepper were mostly Hindus, but the initial purchasers were St. Thomas (Nestorian) Christians. They, however, were too afraid of the Portuguese and their Inquisition to bring the pepper to the coast, so they usually sold it to the Muslim traders who traditionally handled this trade. The Muslims sold it to the Portuguese in Cochin and other coastal forts and also to other exporters. Apparently later in the century some of the Christians were persuaded to bring the pepper themselves, but at least for the first half of the century the Portuguese were largely dependent for their pepper cargoes on their supposed inveterate enemies, the Muslims. These Muslims often rubbed salt in the wound by supplying pepper that was dirty or wet.[45]

That the Portuguese got poor-quality pepper was, however, due more to their own bad management than to the malice of Muslim traders. Ideally, the capital to buy pepper should have arrived on the ships from Portugal late in the year. Since the pepper was harvested in December or January, the Portuguese could have used their capital to make purchases at leisure throughout the next year, buying only clean dry pepper at times when the market was low. About fifteen months after the capital had arrived, the pepper would be loaded on the next year's ships and sent off to Portugal. In fact the Portuguese were never able to get this far ahead. The capital arrived late in the year, and the Portuguese had then frantically to buy whatever pepper was available, get it loaded, and dispatch the ships in time for them to round the Cape of Good Hope before the end of the monsoon. A succession of letters to the king throughout the century described this unsatisfactory arrangement and urged him to have the capital available for the whole year, but the king could never afford to do this. He had to have his capital, in the form of pepper, back in Lisbon as soon as possible. As a result, his officials were reduced to borrowing money from local Portuguese merchants, the city of Cochin, and the raja of Craganore.[46] When the capital arrived, a hard-pressed governor would often take part of it for military necessities despite the king's frequent prohibitions of such conduct.[47]

One final factor should be mentioned which again decreased the control of the Portuguese crown over the pepper production of Malabar : the tendency of both Portuguese officials and private Portuguese merchants to indulge in the highly profitable "illegal" trade themselves. In 1519 the governor left for a cruise in the Red Sea. Seizing their chance, the remaining officials in India reportedly sent cargoes of pepper north to Cambay in ships belonging to the "pirates" of Malabar.[48] A year later a native official of the raja of Cochin told the king of Portugal that Portuguese officials bought pepper in Cochin for their own illegal private trade at prices higher than the fixed official price at which it was bought for the king. Now the local merchants wanted to sell all their pepper at this high price.[49] Other reports of Portuguese and local inhabitants of Portuguese areas trading in pepper, both north to Hurmuz and Cambay, and in southeast Asia, continue throughout the century and later.[50] In the 1630s even clerics, including one of the inquisitors, were accused of such trade.[51]

Once they had the pepper, the local traders who chose to ignore Portuguese trade restrictions had two export routes available : overland to Coromandel and north by sea to Gujarat. The former trade was safe from Portuguese interference, as they had no control over the inland areas of Malabar. Great amounts were transported either by coolie or on bullocks to Coromandel and from there they were distributed all over eastern India and Bengal. Considerable quantities of this pepper were also taken, especially by Gujaratis, to the Red Sea.[52]

The other route, by sea to Gujarat, was considerably more risky. It was navigated in small, light, oared craft called *paroes*, which could escape up the numerous rivers of the western coast if they were seen by the Portuguese. The Portuguese ships were usually too large to follow them, and it was not always possible to land a party to pursue them or carry out reprisals on their helpers. These traders in fact in the 1520s felt themselves so secure that they would sail north to Gujarat and on the way take mocking potshots at the Portuguese coastal forts. They even sailed tantalizingly close to the cumbersome warships of the dreaded Vasco da Gama, when he was back in India as viceroy in 1524, keeping just out of range and showing complete contempt for his armada. Da Gama, who had a tendency to choler at any time, and was certainly more used to massacring Malabaris than to being laughed at by them, "was scandalized at such disrespect".[53] But the Portuguese were far from being totally ineffective. There are frequent reports of their destroying or at least blockading large numbers of Malabari pepper ships. As one example, in early 1525 they caught fifty-three such ships getting ready to sail and burnt them all.[54]

These Malabari *paroes* sailed in fleets, loaded with pepper and bound for Gujarat, of at times over one hundred sail. The pepper they carried was usually consumed locally in Gujarat and North India. Some was taken to the Red Sea, but as Portuguese patrolling across the mouth of the Gulf of Cambay increased in efficiency, once Diu (1535) and Daman (1559) were taken, the Red Sea trade in pepper from Gujarat slackened. It also seems that this tighter control slightly reduced the amount of pepper brought north by the Malabaris ; presumably the overland route to Coromandel was substituted. Certainly the pepper which used to go to Gujarat did not later in the century go to the Portuguese, for they took less and less to Europe as the century progressed.[55]

Other Malabaris were not traders resisting an unjust Portuguese attempt to impose a monopoly—they were pirates. Indeed, piracy had been quite widespread in western India in the fifteenth century, both in the Malabar and Gujarat areas.[56] In the next century piracy continued, and it was not confined to western India. The Persian Gulf, for example, was notoriously dangerous.[57] In the Malabar area open piracy—ships attacking on sight all merchant vessels for the purpose of robbing them—continued in the sixteenth century, though it is clear that some of this piracy was created by the Portuguese. Not only did their monopolistic policies leave little outlet for legitimate (in Indian eyes) trade, but the Portuguese abused those trying to conform to their system.[58] The most flagrant case of this abuse was an important Cochin merchant who traded to Gujarat with two ships. Although he took Portuguese passes each time, some plunder-hungry Portuguese seized his ships. He then moved to the hostile city of Calicut and became an inveterate enemy of the Portuguese.[59] Piracy, in fact, was often resorted to in despair.

Others called "cossarios" by the Portuguese were really guerrilla auxiliaries of the zamorins of Calicut. In the first quarter of the sixteenth century these guerrillas often worked hand in glove with the Calicut merchants whom the Portuguese were trying to displace. The merchants financed their ships in return for protection for the merchants' trade ships.[60] Battles between Portuguese ships and those commanded by members of the Kunjali Marakkar family were frequent in the sixteenth century and this conflict surely was war, not piracy, for the Kunjalis usually worked closely with the zamorins.[61]

In these battles the Portuguese were worsted as often as not. A Jesuit in 1575 said the Malabaris were "absolute rulers of all the seas of India because of our sins and carelessness, and their power and force is waxing while ours is waning".[62] Portuguese ships were afraid to sail alone; another Jesuit said that because of the Malabaris the voyage from Goa to Cochin was more dangerous than that to Portugal.[63] In the early 1580s a letter noted with surprise that the Portuguese fleets had made the western Indian coastline secure against the Malabaris—"an unusual situation".[64] This amelioration was only temporary. In the later sixteenth century all Portuguese ships, and native ships trading within their system, sailed in convoys protected by warships.

Both sides suffered in the continuing Malabar-Portugal warfare, as also did numerous noncombatants, both Portuguese and local. The Portuguese regularly cut down palm trees in enemy areas. This measure was something of a sixteenth century ultimate weapon, for the tree provided several staple products and it took seven years for a newly planted palm to reach maturity. Most Portuguese captains disdained taking prisoners when they captured a Malabari vessel.[65] On one particularly bloody occasion in 1534 a Portuguese captain returned with his fleet to the city of Goa having captured some Malabari ships. The Portuguese fleet sailed in "flying as flags in their masts and spars many hanging Moors, in order to terrorize the enemy and delight the city ; and in order that the Kanarese sailors of this fleet might share in the glory, they took the heads of thirty dead Malabaris and twelve Moors who still lived. They were handed over to the fury of the lads of Goa, who stoned them to death".[66] The Malabaris appear to have responded in kind.[67] A list of Franciscan martyrs show six killed by Malabaris in the 1550s alone.[68]

Given the large and rather clumsy ships the Portuguese used in the first-half of the century, the type of warfare adopted by the zamorin's sailors was very suitable. In 1538 a Portuguese captain told the king how well armed the Malabari *paroes* were and how "150 of them leave from one port and do all the damage they can, and I go after them, and as many again leave from another port and go somewhere else, and do as much damage also, and a man doesn't know which way to turn".[69] These *paroes* usually returned to shore each night, and they maintained close contact with the people ashore. They posted observers on hills who lit fires when a Portuguese fleet was sighted, so warning the *paroes* at sea. The *paroes* then lowered their masts and rowed up a river until the Portuguese had passed.[70] Later in the century the Portuguese finally built lighter ships called *sanguiceis*.[71] This measure improved the effectiveness of Portuguese patrolling, yet they were at times outwitted still. In 1568 a Portuguese ship closed with an enemy and the Portuguese all gallantly boarded her. The Malabaris deserted their own ship, swam to the empty Portuguese one, and sailed happily away.[72]

Many others whom the Portuguese indiscriminately called pirates were in fact legitimate traders. To the Portuguese, anyone trading in pepper, or in anything to Calicut in time of war, was a pirate—but these traders were pirates only from the Portuguese viewpoint. The confusion is obvious in the case of those trading in pepper to Gujarat. It is present also in the following account : In 1525 a Portuguese governor attacked "more than 100 war-paroes of the Malabaris which were coming back from Gujarat, whence they had gone loaded with pepper, and they were bringing back rice and other foodstuffs".[73]

Nevertheless, the distinction between a pirate and a guerrilla fighter or a legitimate trader *was* often blurred. The Portuguese policy of indiscriminate attacks was thus to some extent justified. At times the zamorins disclaimed all responsibility for attacks on the Portuguese, though they did so more because they were at these times threatened by the Portuguese than because the attacks were really carried out without their approval. But the Kunjali Marakkars do seem at times to have operated rather independently, notably in the 1590s.[74] Some traders were not above a little piracy if opportunity offered : in 1530-1531 "some of the subjects of the Zamorin and others . . . set sail in a fleet of thirty grabs for Gujarat, the object of their voyage being generally trade".[75] The operative word is "generally", and the subsequent account of their activities sounds suspiciously like piracy. A fair summing up would be as variegated and perhaps as confused as the sixteenth century situation it described. It would have to admit the existence of considerable piracy in Malabar in the sixteenth century. But it would also have to note that much of this piracy was created by the policies of the Portuguese, while many others whom the Portuguese attacked as pirates were, except in Portuguese eyes, legitimate traders or guerrilla fighters defending Calicut from Portuguese attacks.[76]

The truth, in fact, was a good deal more complex than can be revealed if all seafaring Malabaris are lumped together, whether as corsairs or patriots. Similarly, to describe as corruption all official conduct not fitting an ideal twentieth century model is to obscure rather than elucidate. To follow the categorizations of the sixteenth century Portuguese, whether in acceptance or rejection, is to allow oneself to be blinkered by their point of view, distorted as it was in the midst of battle. They indiscriminately saw Malabaris as hostile and outside the system they had set up to control western Indian trade. But the Malabaris did not accept Portuguese pretensions:

neither should we. Portuguese officials knew nothing of twentieth century standards of official conduct : neither should we when we write of these men. Both "corruption" and "corsairs" were complex phenomena. This is clearly revealed when they are seen functioning in their own time and culture.

NOTES

1. This paragraph was prompted by J.H. Hexter's observations in *Reappraisals in History* (New York, 1963), especially, pp. 201-207.
2. This is not the case for Southeast Asian historiography. For a bibliography of the extensive debate following on from J.C. Van Leur, see David Joel Steinberg, ed., *In Search of Southeast Asia : A Modern History* (New York, 1971), pp. 441-442. For the Middle East, see the recent very provocative study by Edward W. Said, *Orientalism* (London, 1978), a book which must make all western scholars writing on Asian matters rethink their positions. Two excellent studies of the Portuguese chronicles are J.B Harrison, "Five Portuguese Historians," in *Historians of India, Pakistan and Ceylon*, ed. C.H. Philips (London, 1961), pp. 155-169, and C.R. Boxer, *Three Historians of Portuguese Asia : Barros, Couto and Bocarro* (Macao, 1948).
3. Three exceptions I know of are Robert I. Crane, *A History of South Asia* (Washington, 1973), especially pp. 5-27; K.W. Goonewardena "Dutch Historical Writing on South Asia", in Philips, *Historians*, pp. 170-182 and especially pp. 180-182 (both these are rather brief in their discussion of perspective); and B.S. Cohn's provocative "African Models and Indian Histories", in *Realm and Region in Traditional India*, ed. Richard Fox (Durham, N.C. 1977).
4. F.C. Danvers, *The Portuguese in India* (London, 1894), I, xxxix.
5. R.S. Whiteway, *The Rise of Portuguese Power in India* (London, 1899), pp. 174, 324-325; W.W. Hunter, *A History of British India*, 2 vols. (London, 1899-1900). I, 176-185; Vincent A. Smith, *The Oxford History of India* (Oxford, 1919). p. 335.
6. Whiteway, p. 13; Danvers. I, xxviii; Smith, p. 335.
7. F.C. Lane, "National Wealth and Protection Costs", in *War as a Social Institution : The Historian's Perspective*. ed. Jesse D. Clarkson and Thomas C. Cochran (New York, 1941), pp. 39-41; F.C. Lane, "The Mediterranean Spice Trade : Further Evidence of its Revival in the Sixteenth Century", *American Historical Review* XLV (1940), p. 589. Few historians today would argue with C.R. Boxer's explanation of the Portuguese decline. He sees it as being primarily a result of the arrival of the better-organized and better-financed Dutch and English in Asia from late in the sixteenth century. See *The Portuguese Seaborne Empire* (London, 1969), pp. 106-119. See also the stimulating analysis in

Niels Steensgaard's important book *The Asian Trade Revolution of the Seventeenth Century : The East India Companies and the Decline of the Caravan Trade* (Chicago, 1974). In his view the Portuguese had no chance of success, once they were faced with the competition of the Dutch and British companies, because of the "structural differences" between the Portuguese and the caravan traders on the one hand and the Dutch and the British trading companies on the other.

8. Whiteway, *Rise of Portuguese Power*, pp. 13, 124 fn.
9. Hunter, *History*, I, 317.
10. Herbert Butterfield, *The Whig Interpretation of History* (London, 1931), p.v. More recently Sylvia Thrupp has coined the rather unlovely term "temporal ethnocentrism" to describe basically the same phenomenon. J.H. Hexter reverses the coin in his amusing article "The Historian and His Day" (*Reappraisals*, pp. 1-13), for he claims to know more about early modern Europe than about present day America and unconsciously takes the perspective of the former when looking at the latter. This is perhaps an ideal perspective, but for most of us it will remain only an ideal.
11. Max Weber saw "legal domination" as the essential element in a modern state and regarded the development of a modern bureaucracy as necessary for legal domination to become effective. Several specific characteristics have to be found in a bureaucracy in order for there to exist "a system of legal domination in which the exercise of authority consists in the implementation of enacted norms". For these see Reinhard Bendix, *Max Weber : An Intellectual Portrait* (New York, 1962), p. 424. For a translation of Weber's own words see H.H. Gerth and C. Wright Mills, *From Max Weber : Essays in Sociology* (New York, 1958), pp. 196-204. For an important discussion of the conflict between the ideal of the modern bureaucratic state and the survival of earlier practices in modern Southeast Asia see W.F. Wertheim, "Sociological Aspects of Corruption in Southeast Asia", in *State and Society : A Reader in Comparative Political Sociology* ed. Reinhard Bendix (Boston, 1968), pp. 561-580. Gunnar Myrdal finds a similar situation prevailing in India to-day : *Asian Drama* (New York, 1968), II, pp. 948-950. See also Ralph Braibanti, "Reflections on Bureaucratic Corruption," *Public Administration* XL (1962), 357-372. Steensgaard is again interesting on this topic : *Asian Trade Revolution*, especially pp. 19-21, 93-95.
12. Gerth and Mills, *From Max Weber*, p. 224. and generally pp. 196-244.
13. For the scope of these legal perquisites in the 1580s see B. Mus., Add. MS 28433, ff. 19-74v and 166-173, where are set out the various voyages and other privileges for the captains of most of the forts of India.
14. AHU, "India", caixa 5, account of 15 February 1618.
15. Diogo do Couto, *Dialógo do Soldado Prático* (Lisbon, 1937), p. 14.

16. For example, LMBP, III, 167-198; LM, XII, 44v-45; LM, XXVIIB, 378-8v.
17. Alberto Iria, *Da Navegação Portuguesa no Indico no século XVII* (Lisbon, 1963), p. 84.
18. APOCR, VI, 1250-1251; see also ANTT, "Colecção de São Vicente", XXVI, 93v-94v.
19. José Wicki, "Duas relaçoes sobre a situação da India Portuguesa nos anos 1568 e 1569", *Studia* (1961), 177-178; *Studia* (1959), 75-76; "Provisoes dos Vice Reis", HAG, II, pp. 129-129v; ANTTCC, 1-97-26; APOCR, III, 105-106, 564-566; AHU, "India", caixa 14, petitions of September to December 1642; Gavetas, V, 327.
20. Couto, VIII, cap. 1.
21. APOCR, III, 230-292; "Provisoes, Alváras c Regimentos", HAG, II, 276v-277; "Livro Morato", HAG; ff. 253-254.
22. Castanheda, VI, xxxv; Correa, II, 508; ANTTCC, 1-92-88.
23. Gavetas, III, 213.
24. ANTT, "Colecção de São Lourenço", IV, 509. For similar conduct in early seventeenth century Ceylon see P. Pieris, *Ceylon and the Portuguese, 1505-1658* (Tellippalai, Ceylon, 1920), p. 191.
25. ANTTCC, 1-108-12.
26. Couto, XII, i, 8. See his *Dialógo do Soldado Prático* for detailed and trenchant criticism of the prevailing abuses.
27. Gavetas, V, 327.
28. A.R. Disney *Twilight of the Pepper Empire* (Cambridge, 1978), p. 61, and generally pp. 56-62; Niels Steensgaard, *The Asian Trade Revolution*, pp. 12-113. The quotation is from p. 95. As is to be expected, the situation in premodern China was similar : "In the Chinese empire the term 'corruption' must be understood in a special sense as the carrying to excess of practices that, when not excessive, were a normal and recognised part of the government". J.K. Fairbank, E.O. Reischauer, and A.M. Craig, *East Asia : The Modern Transformation* (Boston, 1965), p. 108. Similarly, in Iran until very recently "the derivation of the largest possible personal profit from government office was not only expected but was regarded as his official right. . . . The practice was no secret, nor did any stigma attach to it. It was only later that the Western attitude toward peculation by public officials came under discussion and ostensibly won public approval". Joseph M. Upton, *The History of Modern Iran: An Interpretation* (Cambridge, 1968), pp. 23-24.
29. See Couto, VIII; cap. 19; LM, XII, 261; ANTT, "Colecção de São Lourenço", I, 338-338v; António Bocarro, *Década 13 da História da India* (Lisbon, 1876), p. 213; Diogo do Couto, *Vida de D. Paulo de Lima Pereira* (Lisbon, 1903), pp. 23, 47, 72, 76-77; Ludovico di Varthema, *The Itinerary of Ludovico di Varthema* (London, 1928), p. 63; ANTT, Colecção de São Vicente", XIV, iv; A. de S.S. Costa Lobo, ed., *Memórias de um soldado da India* by F.R. Silveira (Lisbon, 1877), pp. 83-84.

30. For example, A. Botelho de Sousa, *Subsídios para a história militar maritíma da India* (Lisbon, 1930-1956), I, 108-124, 353-357; II, 16-17, 243, 357.
31. Respectively, in C.R. Boxer and Frazão de Vasconcelos, *André Furtado de Mendonça* (Lisbon, 1955), pp. 11, 21; "The Portuguese in the East (1500-1800)," in *Portugal and Brazil: An Introduction*, ed. H.V. Livermore (Oxford, 1953), p. 233. *The Portuguese Seaborne Empire* (London, 1969), pp. 58, 298, 299.
32. O.K. Nambiar, *Portuguese Pirates and Indian Seamen* (Bombay, 1955); *The Kunjalis: Admirals of Calicut* (London, 1963).
33. K.M. Panikkar, *Malabar and the Portuguese* (Bombay, 1929), p. 143.
34. *Boletim do Conselho Ultramarino : Legislação Antiga* (Lisbon, 1867), I, 59. Similar prohibitions throughout the century and later are given in APOCR, V, 292; III, 6-7, 567-568, 650-652; "Livro Morato", HAG, ff. 61-62; "Livro Verde", HAG, I, 101v-103v; LM, XIXD, 1200v.
35. *Boletim do Conselho Ultramarino*, I, 59.
36. *Annaes Maritimose Coloniaes* (Lisbon, 1840-1846), IV, 120; ANTTCC, 1-27-78; ANTT, "Colecção de São Vicente", XIV, 161; ANTTDR, VIII, 183.
37. Recent discussions of this important topic include ; Jan Kieniewicz, "The Portuguese Factory and Trade in Pepper in Malabar during the 16th Century", *Indian Economic and Social History Review* VI (March 1969), pp. 61-84; Vitorino Magalhaes-Godinho, *L'Economie de l'Empire Portugais aux XVe et XVIe Siècles* (Paris, 1969), pp. 537-828; M.N. Pearson, "Commerce and Compulsion : Gujarati Merchants and the Portuguese System in Western India, 1500-1600", Ph.D dissertation, University of Michigan, 1971, pp. 106-129; A.R. Disney, *Twilight of the Pepper Empire*, pp. 1-49.
38. Barros, I, vi, 5.
39. Couto, *Vida de D. Paulo de Lima Pereira*, pp. 76-77; Zain-al-Din, *Tohfut-ul-Mujahideen*, trans. M.J. Rowlandson (London, 1833), pp.153-154; Francisco de Andrade, *Chrônica do muyto alto e muyto poderoso Rey destes Reinos de Portugal Dom João o III deste nome* (Coímbra, 1796), IV, 499-500; Barros, I, vi, 3; Couto, VII, x, 17.
40. *Cartas de Affonso de Albuquerque* (Lisbon, 1884-1935), I, 126.
41. Castanheda, VI, lvii; VII, xii, xliii; *Orient Português* (Goa) II, 286.
42. See, for example, ANTT, "Colecção de São Lourenço", III, 34, 305v, ANTTCC, 1-107-9; Couto, VII, x, 3; DI, VII 88-89, 530; LMBP, IV, 312.
43. *Cartas de Affonso de Albuquerque*, IV, 174-175. This source on pp. 175-176 estimates the total production of the Malabar area at about 16,000 bares. Some 2,000 to 2,500 bares were used locally, 3,000 escaped inland and 500 to 600 went to Diu by sea. Of the remaining 9,000 bares, the Portuguese got some, but a large amount also went to Hurmuz, Diu, the Red Sea, and Coromandel. See also DUP, III, 373-5. A report of 1533 explained to the king why he got so little pepper. Other merchants paid "all in money and at the requested

price, and Your Highness pays in copper and money and at a fixed price" (ANTTCC, 1-50-65). See also José Wicki. ed., "Duas Relaçoes sobre a situação da India Portuguesa nos anos 1568 e 1569", *Studia* (1961), 216-217, and *Cartas de Affonso de Albuquerque*, I, 329-330.

44. Couto, VI viii, 2, 3, 4, 8, 9; VI, ix, 18, 19; VI, x, 15, 17. For the pensions see P.S.S. Pissurlencar, ed., *Regimentos das Fortalezas da India* (Bastorá, Goa, 1951), pp. 217-219; "Alvarás e Provisoes de Sua Magestade", HAG, I, 77; DUP, I, 261-2.
45. "Cartas de Vice Reis", ANTT, 95; António da Silva Rego, *História das Missões do Padroado Português do Oriente* (Lisbon, 1949), p. 390; Antonio da Silva Rego, ed., *Documentação para a história das missões do Padroado Português do Oriente : India* (Lisbon, 1947), II, 353-354; IV, 479-480; Castanheda, VI, lxxiii.
46. A. da Silva Rego, *História*; p. 139: A. da Silva Rego, ed., *Documentação* IV, 495; Barros, III, iv, 7; Artur Basílio de Sá, ed., *Documentação, para a história das missões do padroado Português do Oriente : Insulíndia* (Lisbon, 1954) I, 333-334; LM, vol XXIA, 222, ANTTCC, 1-30-36, 1-52-32, 1-82-2, 1-86-89, 1-94-73, 1-107-86.
47. For example, "Alvarás e Provisoes de Sua Magestade", HAG, I, 78-78v. 110. 202v.
48. ANTTCC, 1-25-108.
49. ANTTCC, 1-27-69.
50. ANTTCC, 1-83-74, 1-94-54; AHU, caixa 3, no, 4; Gavetas, III, 215.
51. LM, vol. XIIIA, 220-220v; XVII, 68,
52. *Cartas de Affonso de Albuquerque*, IV, 175; ANTTC, 1-30-36, 1-105-90; LMBP, II, 156.
53. Barros, III, ix, 2; ANTTCC, 1-30-36.
54. Barros, III, ix, 5.
55. ANTT, "Colecção de São Vicente", XI, 37; DUP, I, 418; Barros, III, ix, 2-3; III, ix; 5-6; III, x, 9; IV, iv, 17; J. Wicki, ed., "Duas relaçoes", pp. 210-211.
56. S.D. Goitein, *Studies in Islamic History and Institutions* (Leiden, 1966), p. 348; ANTT, "Colecção de São Vicente" XIV, iv; "Ma' athir-i Mahmud Shahi", corrected copy in Department of History, M.S. University of Baroda, ff. 21-23; S.C. Misra and M.L. Rahman, eds., *The Mirat-i Sikandari* (Baroda, 1961), pp. 131, 145-146; John Briggs, trans., *The Rise of the Muhammadan Power in India* (Calcutta, 1908-1910), IV, 65 for pirates off Gujarat; for Malabar, Abd-er Razzak's account in R.H. Major, ed. *India in the Fifteenth Century* (London, 1857), p. 18; *The Itinerary of Ludovico di Varthema*, p. 63.
57. Bibliotheca Nacional de Lisboa, "Descrição das terras da India Oriental", MS. 9163, f. 23; ANTTCC, 1-86-120; Couto, X, v, 7; for the Persian Gulf see ANTTCC, 1-35-52; ANTT, "Colecção de São Lourenço", II, 30; IV, 459v; ANTT, "Livro de Leis de D. Manual", f. 137.
58. Correa, II, 519; Castanheda, VI, xlviii; Zain-al-Din, pp. 157-158.

59. Barros, IV, viii, 12.
60. Correa, II, 777; Barros, III, ix..3.
61. Barros, IV, vi, 25; IV, viii, 12; ANTTCC, 1-77-36; Boxer and Vasconcelos, *André Furtado de Mendonça*, pp. 11, 21; ANTT, "Colecção de São Vicente", V, 132v; R.S. Whiteway, *The Rise of Portuguese Power in India*, p. 47; DUP, I, 422; Correa, II, 777, 810; APOCR, III, 284.
62. DI X, 3,
63. Ibid., VIII, 120-21.
64. A. da Silva Rego, *Documentação*, XII, 783.
65. Diogo do Couto, *Vida de D. Paulo de Lima Pereira*, pp. 76, 77; Couto, VII, x, 17; DUP, I, 418.
66. DUP, I, 422:.
67. For example, Couto, VIII, cap. 19; APOCR, III, 928; DI, VI, 278.
68. *O Oriente Português*, VIII, 22-26.
69. ANTT, "Colecção de São Lourenço", I, 338v.
70. J. Wicki, ed., "Duas relaçoes", p, 211; A de S.S. Costa Lobo, ed., *Memórias*, pp. 83-84.
71. APOCR, III, 929 ; Couto, XII, ii, 2. On the etymology of this word see Couto, X, v, 7, and J.H. van Linschoten, *The Voyage of John Huyghen van Linschoten to the East Indies* (London, 1885), II, 169-170.
72. Couto, VIII, cap. 28.
73. Castanheda, VI, lxxxiii,
74. Couto, VII, x, 17; Linschoten, II, 169-170; F. Mendes da Luz, *O Conselho da India* (Lisbon, 1952), pp. 440-441.
75. Zain-al-Din, p. 126.
76. O.K. Nambiar, *The Kunjalis*, p. 97; *The Travels of Pietro Della Valle* (London, 1892), II, 201, 356; S.N. Sen, ed., *Indian Travels of Thevenot and Careri* (New Delhi, 1949), pp. 18-19; LM, XIV, 214-5. To at least some extent R.S. Whiteway did try to discriminate : "Some of the so-called pirate leaders were, however, commanders under the Samuri, carrying on a guerrilla war". (p. 47). W.H. Moreland called them "pirates", who "treated the Portuguese ships precisely as the Portuguese treated theirs" but spoiled it by saying they were "Arab 'pirates'". *India at the Death of Akbar* (London, 1920), pp. 202, 206, 210. For interesting parallels with the situation in nineteenth century Malaya see Nicholas Tarling, *Piracy and Politics in the Malay World* (Melbourne, 1963). It may also be noted that Fernand Braudel's depiction of piracy in the Mediterranean in the sixteenth century contains many interesting parallels with what I have said in this chapter. See *The Mediterranean and the Mediterranean World in the Age of Philip II* (London, 1972-3, 2 vols), II, 865-91.

THE CROWD IN PORTUGUESE INDIA 3

OUR understanding of overt sub-élite political activity in pre-industrial times has increased enormously as a result of the pioneering work of George Rudé.[1] Thanks to him it is no longer possible to see crowds in pre-industrial Europe as undifferentiated, lumpish, violent, arbitrary mobs. What follows is an attempt to extend these European studies into an earlier, and a colonial, setting; this chapter discusses sub-élite political activity in the Portuguese Asian empire in the sixteenth century. Three background points must be made clear here. First, most of the data come from the Portuguese capital in the East, the city of Goa, which was conquered by Afonso Albuquerque in 1510. Nevertheless, some case studies relate to such dispersed Portuguese forts and settlements as Ternate in modern eastern Indonesia, Cochin in southwest India, and Diu and Daman in Gujarat. Second, this chapter is concerned almost exclusively with the political behaviour of Portuguese and part-Portuguese people in this empire. This is not done for ethnocentric or neo-colonialist reasons, but simply because this division reflects accurately divisions in contemporary Portuguese colonial society. Indigenous inhabitants of Portuguese areas were certainly economically powerful and politically potent,[2] but their political activities did not impinge on those of the lower-order Portuguese. Third, this study is not exclusively concerned with crowds and violent mobs. Some attention will be given to less open and less forceful political activity.

Portuguese colonial society was divided according to three different, though overlapping, criteria; "purity" of blood, rank, and marital status. The first was the familiar Iberian gradation, starting with those born in Portugal (*reinoes*) and descending through those born in Asia of Portuguese parents (*castiços* or *Indiaticos*) to those born in Asia of Portuguese and Indian, or Portuguese and African, parentage (respectively, *mestiços* and mulattoes). Outside of this ranking, and beneath it, were the native Christians,[3] but even the castiços and the mestiços were regarded with considerable suspicion by the élite. Thus one reads of Nicolao d'Abreu de Mello, "who, although born here, is the son of Portuguese parents and a well-conducted person".[4] The received view on castiços was that they were "usually ill-behaved and not to be trusted".[5] Mestiços were even most suspect. Late in the 1540s, an experienced official told the king that seven new men had recently been hired for clerical positions in a government office; "Of the seven, four are mestiços, something which absolutely astonished me". In 1634, the viceroy told the king that Portuguese men were needed in Diu, "because most of the soldiers there are almost niggers".[6]

The second categorization is equally familiar—the division of the Portuguese population into the three estates of ecclesiastics, nobility, and people. This division seems to have been most visible on ritual occasions, such as when the oath of allegiance to a new king was being given.[7] The third criterion was the simple division between unmarried men from Portugal, who were called *soldados* and were liable to military service, and those Portuguese or part-Portuguese who had married and settled in Asia, the *casados*.[8] The Portuguese affected by this division were primarily from the third estate, the people. It obviously did not apply to the first estate, while the "nobility", whether married or not, were usually liable and eager to engage in military service until incapacitated by illness or old age.

These divisions were recognized by contemporary authors, but they were not completely watertight. In practice a mestiço could be a de facto noble, and a casado could be a soldier, especially in an emergency. More important, in times of peace these divisions appear to have been subsumed in a common Lusitanian equality, an equality that was created by the basic fact that this was a colonial society.

The Portuguese Asian empire consisted of a string of forts on the littoral of Asia. Each settlement was surrounded by a mass of foreign, strange, and frequently hostile native populations. In India these neighbours were often subjects of powerful states : Goa was surrounded by the strong Muslim state of Bijapur, and Diu and Daman by the sultanate of Gujarat and, after 1572, by the Mughal empire, arguably the strongest and richest state in the world in 1600. In the face of these states, the Portuguese were not always united, nor did they form a cohesive group, yet clearly a sense of Portuguese identity was stronger here than in the motherland. Hence, contemporary accounts which say that among the Portuguese in Goa "there one is no better than another, as they think, the rich and the poore man all one, without any difference in their conversations, curtesies and companies".[9] Hence also the customary wearing of swords by all Portuguese except priests, doctors, and lawyers, and the likening of the daily auction on Goa's main street, the Rua Direita, to "the bourse at Antwerp, except that 'hether in Goa there come as well Gentlemen, as marchants and others. . . .' "[10].

This sort of unity appeared most fully in conflictless situations, such as the more important religious processions and occasions of public rejoicing. Here all Portuguese could unite, at least for the day, and celebrate their common nationality, despite their location 10,000 miles and three months sailing distance from Portugal. It was not coincidental that these occasions usually celebrated Portuguese victories over native armies, or else were specifically religious and thus exclusive of the non-Christian population. The best example of the former was D. João de Castro's huge, bacchanalian, and self-congratulatory triumph in April 1547, to celebrate his relief of the siege of Diu in 1546.[11] Another prominent occasion, the auto da fé, called together a crowd for a religious activity. These were held every one or two years; the cathedral bell tolled to summon the Christian population to witness the procession of people relaxed by the Inquisition to the secular authorities. Two large special public celebrations were associated with the great Jesuit missionary, Francis Xavier. Both the arrival of his body in Goa in 1554, and his canonization in 1642, were occasions for massive public rejoicing. Particular saints' days, notably those of St. Paul and St. Catharine, and other religious holidays such a Corpus Christi, were celebrated each year with great éclat by the entire Christian population. The workers' guilds took a prominent part in these. Thus the dockyard workers provided a ship or galley for the procession, the tanners a castle, the cobblers a dragon, and so on. Others impersonated or depicted historical or mythological personages: King David, various other kings, emperors, princes, devils, giants, wizards, and ladies of easy virtue.[12]

Such fiestas tell us little about lower-level political activities, or groupings, for these obviously became clearly apparent only in situations of stress or conflict. Two groups within the third estate, with different interests, life-styles, and methods of solving problems need to be distinguished. The settled, married, commercially-minded casados were very different from the soldiers, despite the fact that the former were recruited from the ranks of the latter.

The size of the Portuguese soldier population in any fort depended on the season. It was greatest during the period of the southwest monsoon, from June to September, for during this time the strong off-shore winds precluded any patrolling along the western Indian coastline. Up to 5,000 footloose and idle soldiers could be resident in Goa in this rainy season. During the rest of the year while the many fleets of small warships were out opposing "piracy" and checking native ships for "contraband cargoes", the soldier population of Goa fell to a few hundred.[13] This, however, changed toward the end of the year when the annual fleet from Portugal reached Goa. In the sixteenth and early seventeenth centuries, about 1,000 men came on each fleet to replenish the Indian garrisons.[14] These men remained in Goa, independent and subject to no discipline, until they were recruited by a captain, whom they would then follow to his assignment.

These soldiers were considerably brutalized even before they reached Goa. Some of them were beggars, vagrants, and jail-birds who had been forcibly impressed in Portugal. Increasingly, soldiers were in fact convicts who had been sentenced to serve in India rather than in jail in Portugal.[15] Nor were they refined during the long voyage to India. Food was scanty and became, as the voyage progressed, increasingly putrid and stale. Disease and mortality were ever present. Of 5,228 soldiers who left Lisbon in 1629-34, only 2,495 reached Goa; the rest succumbed to sickness, deserted in Lisbon, or perished in shipwrecks.[16] Nor did arrival in Goa mark the end of the threat of disease; cholera and other maladies related to impure water were endemic.[17] In thirty years, from 1604 to 1634, it is claimed that an incredible total of 25,000 soldiers died in a single facility in Goa, the Royal Hospital.[18] In addition to health problems, soldiers experienced the brutality and contempt of their superiors and other detrimental conditions of the unstable frontier-type of society of Goa, on the edge of a large, hostile, and strange subcontinent.[19]

During the rainy season of June to September, the soldiers usually gathered in groups of about ten to a lodging. Their lack of money and occupation soon caused a restlessness which often led to violence. It was "somewhat dangerous" to walk about Goa after nine at night.[20] A Jesuit account of the rainy season in Cochin in 1569 starts by saying the soldiers, "lived quietly and peacefully and did not cause many scandals", but goes on to mention several knifings, beatings, and sundry other violence.[21] Another illustration of the brutish nature of life in Goa was a soldiers' riot in 1548. The Portuguese governor provided a meal each day for several hundred indigent soldiers. Despite this generosity, the soldiers were so "low and shameless" that their dining room became a disorderly place filled with shouting and brawling soldiers. An official who castigated them and tried to restore order was assaulted for his pains; there ensued a full-scale battle between the soldiers and the governor's guard.[22]

Given this ambience, the prevalence of riots, rebellions, and general disorder among the soldiers is hardly to be wondered at. These activities took two forms. Sometimes the soldiers rioted on behalf of their leaders, the *fidalgos* (a "noble", literally "son of a somebody"), over a matter of concern to the noble rather than to the soldiers. These élite-led mobs will be examined but first it is necessary to examine riots directed *against* commanders.

Historians, themselves usually middle class and seldom given to violence, have usually condemned the pure, non-élite-led crowd as being "directionless", a "mob".[23] In Portuguese Asia these crowds may have been violent and usually unsuccessful, yet in fact they were responding to clear threats to their interests in the only way they could. As such, their behaviour had, on their terms, a rationale. An example was a near-mutiny during the famous second siege of Diu in 1546. This Portuguese fort had been invested for some months by a strong Gujarati army and had been extensively mined. The common soldiers objected to being shut up in a potential deathtrap and demanded a sortie. Wiser counsels from the fidalgos and priests failed, and, faced with imminent rebellion, the élite capitulated and an attack was made. It failed and the fort was nearly lost;[24] yet, given the soldiers' situation and their ignorance of the strength of the Gujarati army, their demand was understandable.

A comparable, but more successful, occasion was a mutiny at sea in 1583. After a long battle between twenty small Portuguese warships and a large hostile trader, the enemy capitulated in return for a guarantee of the lives of those on board. But a surrender was not good news for the Portuguese soldiers and sailors; they preferred to board, waving their swords and shouting "Santiago!". In the resulting confusion all sorts of trifles, such as gold and jewels, could be picked up, and later disposed of in Goa to a discreet Indian merchant. Thwarted of their sack, the Portuguese mutinied, shouted abuse at their captain, took over fourteen of the ships, and raised the black flag. Upon reaching the Portuguese fort of Daman, the nearly 300 mutineers terrorized the town officials and the casados. Subsequently the six loyal ships arrived, and the mutineers attacked them and tried to kill their unsporting captain. A confused battle followed, which even a group of priests, with crucifixes uplifted, were unable to stop. The captain fled to the sanctuary of a church. After negotiations, all the rebels were given a share of the prize money, and the mutiny ended.[25] Here at least a crowd achieved what it saw as its legitimate due.

In these instances the soldiers acted independently of, indeed in opposition to, their leaders. Such cases may tend to point to class divisions in Portuguese colonial society, and indeed divisions of status at least were clearly recognized. Thus a report on the Portuguese defense of their fort at Hurmuz in 1622 listed by name twelve dead—eleven fidalgos and a priest—and added, "And many other ordinary people were killed fighting bravely".[26] Similarly, at an inquiry into a scandal of 1601, a sailor testified that he "didn't want to know anything about it because big people did it".[27] Nevertheless, a much more frequent pattern was violence by a group of soldiers directed by a fidalgo. These crowds were recruited by patronage ties.

A lone soldier, newly arrived in Goa or temporarily resident in a west coast port during the rains, had little choice but to live on a fidalgo's charity. He usually received no pay at such times and had no trade or skill. In any case, until he married he could hardly establish himself with a trade or a house, for he was very much of a sojourner. He had no choice but to eat, and probably live, with a noble or to accept charity from the church. The latter option seems to have been looked down on, and in any case, it did not lead to employment. Thus, in 1548, three or four fidalgos gave money to the soldiers, and the governor fed nearly 800 of them twice a day.[28] In 1609, the French traveller Jean Mocquet was encouraged to emulate all the poor Portuguese soldiers and eat at a noble's table. He speaks of the lords and gentlemen of Goa feeding twenty or thirty each.[29] Indeed, there is even one reference to an important noble having his followers grouped in and around his own house. Behind the Rua Direita "was another street of the houses of Heitor da Silveira".[30] He presumably lived there, surrounded by men housed and fed by him; in essence, his private army controlled this area.

Nor indeed were soldiers the only ones thus bound to a patron. There are scattered references to eminent nobles having as their followers various fidalgo relatives and also men described as *criados,* who in this context are clearly not servants but lower fidalgos and other clients of the important noble, perhaps best described as "attendants".[31] A very precise description of this whole recruitment process comes from João de Barros: Christóvão de Sousa, the captain of the fort of Chaul, was "a fidalgo of high quality, personally energetic and humane, a polished talker, jovial, and friendly with everyone; and he was magnificent not only from the continual food which he provided but also from the monetary assistance that he gave to those who had none; as a result more fidalgos spent the rains in Chaul than in any other part of India; and he had so much authority, and so many followers, any side helped by him had a great advantage".[32]

Here was the normal payoff for the fidalgo who fed soldiers during the rains. Such soldiers would be expected to form part of his entourage as he rode about the streets and when he went to church on feast days.[33] If their noble was insulted, or wanted revenge for some other reason, his diners became his private army.[34] More important, and also more legitimate, the main object of this system was its use as a recruitment device for captains newly appointed to command a ship or a fort.

The soldiers were not under the continuing authority of any captain during the rains; every September they were free agents available to whichever captain offered most or had in the past treated them best. The captains would receive pay for the number of troops sanctioned for their command, "which done the Captaines bidde their soldiers to a banket, and of their own purses give them something besides their pay, for that every Captaine seeketh to have the best soldiers and buy much victuailes and other thinges at their owne charges, thereby to have their soldiers good willes, and to use them well".[35] Clearly a captain who had fed soldiers during all the rains would have little difficulty recruiting them in September.[36]

These dependents could on occasion be mobilized and used by their masters in their own private wars. In 1569, in Cochin, there was a confrontation and nearly a major battle between the local town officials and the citizens on the one hand, and some captains and their soldiers on the other.[37] Perhaps more revealing was a curious incident in Goa in 1601, when the statue of Dom Vasco da Gama, the discoverer of India, was toppled, chopped apart, and distributed around the city : the head and a hand in the town pillory, another hand in the auction mart of the Rua Direita, and so on. The statue had been erected by da Gama's grandson, Dom Francisco da Gama, who had been viceroy in 1597-1600. His authoritarian attempts to reform Portuguese India had aroused considerable antipathy, and this apparently was the motive for the desecration. At an inquiry in 1609, most witnesses denied all knowledge of the event, contenting themselves with expressing routine outrage and horror at such a heinous and scandalous crime. This forgetfulness was possibly aided by da Gama's general unpopularity, and certainly total recall was hindered by a public threat to kill any informer. Nevertheless, a few intrepid witnesses did accuse a fidalgo "and his people" or "and his soldiers" of responsibility. Whatever the motive, it is clear that here common soldiers were used by their patrons in the commission of this act.[38]

The best illustration of the fidalgos' use of their soldier clients in their own political battles came during the disputed succession for the governorship of Portuguese India in 1526-27. Usually such important decisions were settled within the councils of the élite, the first and second estates, and without reference to the opinions of the people. But in this case the élite remained divided, resulting in a long and complicated dispute which involved all the Portuguese forts from Cochin to Chaul.[39] In the sometimes violent debate over a matter which was beyond the horizon of a common soldier, various factions of nobles occasionally would mobilize their supports in order to intimidate the opposition. One heard of fidalgos haranguing their supporters "in their lodgings and at their tables", of fidalgos and "others of their party" meeting, and nobles threatening to lead open revolts. The whole dispute was finally settled peacefully, but it demonstrates how a fidalgo could mobilize his followers for his own political aims. Thus, here again, the crowd's participation was not motiveless. True, they were directed by their betters, but again their participation was a logical act, in this case based on their cliental ties to the élite.

The casados in the various Portuguese settlements had different interests, leaders, and methods of political activity from the soldiers. They had nearly all come out as soldiers to India and had married and settled down. Given the small number of Portuguese-born women in India the casados born locally must nearly always have been mestiços rather than castiços. They were apparently outnumbered by the soldiers,[40] but despite this their influence was greater. First, they were permanently resident in Goa and the other settlements, while the soldiers were only sojourners. Second, rates of mortality among the soldiers, from disease rather than violent death, were very high, much higher than among the more settled and stable casado population.

It is difficult to find out what the casados did in Goa, for the sources usually do not distinguish between a native Christian (who always took a Portuguese name on baptism) and a mestiço or Portuguese. Thus João Fernandes could be a *reinol*, or a native, or somewhere in between. The picture that emerges from many scattered sources is that the casados did not normally do manual work, nor did they engage in retail trade. Rather, they seem to have been wholesalers, owners of property including shops, supervisors of groups of non-Portuguese craftsmen, and above all sea traders.[41] The large slave population in Goa helped the Portuguese avoid manual work, as slaves were frequently hired out by the day or served in their masters' shops. Female slaves were sometimes used as prostitutes by their owners, but whatever the employment, a slave's earnings reverted to the owner.[42]

The extant evidence seems to show that most casados lived by sea trade and by selling wholesale cargoes they imported from all over Asia. This trade began almost as soon as the Portuguese reached Asia. Thus the great conqueror Afonso Albuquerque told the king in 1513 that "Your people travel securely all over India, both by land and sea. . . they buy and sell all over the Malabar area".[43] Ten years later another Portuguese told the king that wood for shipbuilding was now scarce in Cochin, because the local Portuguese were building their own ships, for they intended to settle in India and to live by trade.[44]

The Portuguese seem to have traded with virtually every major port in Asia. Among those mentioned are ports in

Indonesia, South Arabia, Burma, Thailand, Malaya, East Africa, Persia, Japan, Ceylon, and China, as well as in all parts of India.[45] But the most important route sailed was to the great ports in the Gulf of Cambay. The capital involved on this route was greater than any other within Asia.[46] The passage was made in fleets of perhaps 300 or 400 small ships, with two or three fleets making the voyage each year, guarded by Portuguese warships. A Frenchman who sailed with this fleet noted, "There are but few of the inhabitants of Goa, whether Christians or other, that have not a share in this fleet, or at least in some of the ships of Goa and the other ports of the Portuguese".[47]

The crews on Portuguese-owned ships were all local sailors—usually Muslim—except for one or two Portuguese who were the captain, master, or pilot.[48] On longer voyages, presumably sailed in larger ships, more Portuguese were found on board.[49] Further, many Portuguese-owned trade goods were shipped on Indian-owned ships, and another sign of the way in which these casados fitted into the local Asian environment is the way in which they often would decide to settle, either temporarily or permanently, in non-Portuguese trade centres. The most popular of these was the great port of Cambay, in Gujarat at the head of the Gulf of Cambay.

Trade to Cambay had started at least by 1509[50] and continued thereafter. It was carried on despite discouragement from both church and state. The first Provincial Council of 1567 disapproved of Portuguese settling in areas in which there were no priests, as they would seldom be able to take the sacraments. The governors feared the Portuguese residents in Cambay could serve as hostages in the event of a war between Portugal and this Muslim state of Gujarat; on the other hand, if the Goa-Gujarat trade was left to the Gujaratis, *they* would be hostages for Portugal in case of war.[51] But the occasional prohibitions had little effect. When the Mughal emperor Akbar arrived at the port of Cambay in 1572, during the first Mughal conquest of Gujarat, he found fifty or sixty Portuguese, and these, a remnant of the normal population, were the only ones who had been unwilling to leave the war zone, because they had goods in the town.[52] In 1594, there were about 100

Portuguese families in Cambay; the number seems to have decreased in the seventeenth century.[53] Many of these Portuguese traders evidently settled permanently in Cambay and married local women; a Jesuit who baptized 180 people in Bassein in 1573 noted that among them were "many mestiço women and boys, children of Portuguese in Cambay".[54] The main activity of these Portuguese was the purchase of Gujarati goods to send to Goa, where they were re-exported to Europe and all over Asia on Portuguese ships. Some were presumably agents of large Portuguese merchants in Goa who preferred to buy their goods at the source rather than wait until they had been brought to Goa.[55]

In all of this, it is important not to categorize people too strictly. Thus fidalgos could and did own merchant ships or trade in the ships of others. Similarly, the casado population was not a clearly defined group; it shaded off into the bulk of the native population. In this off-white area were many people who claimed Portuguese blood and who worked with their hands. These lower-class Portuguese were organized into guilds modeled on those of medieval Europe.[56]

Higher-class Portuguese were members of other councils, of which the most important was the Council of State. This was entirely an élite body, for apart from one ecclesiastical representative its members were all fidalgos. The casados were thus dependent for their official political voice on the *Senado de Camara*, or Municipal Council. These were established in most of the large Portuguese settlements in Asia. Membership on one of these elected bodies was almost exclusively restricted to married Portuguese male citizens. This, however, meant that not only casados but also resident fidalgos could vote for and sit on the municipal council. Nevertheless, this influence from the *poderosos* was mitigated by the four *procuradores dos mesteres*, or councillors elected by the guilds, and so representative of the lower-class casados.[57]

The activities of the municipal council will be discussed shortly, but it may be noted here that in two ways its influence was less than that of the equivalent bodies in contemporary London and Amsterdam.[58] First, the Câmara was subordinate to the Council of State, and was not expected to concern itself with the high mysteries of diplomacy and war and peace which that council handled. Second, the bulk of the population of Goa, including some of the city's wealthiest merchants, was not represented by the municipal council. This was because only Portuguese or mestiços could vote for the municipal council; in the context of the Portuguese colonial empire such discrimination is hardly to be wondered at, but it did mean the municipal council represented only the white, and thus the less wealthy, part of Goa's non-elite population.[59] Until the late eighteenth century, even local Christians were excluded. This restriction meant that in most matters the indigenous population of Goa governed itself or was subject to decisions by the municipal council over which they had no formal control. In such matters as customs duties, or the *colecta*, a tax on food, they usually were not consulted, at least formally. Normally their only recourse, which they occasionally used, was to vote with their feet and leave the Portuguese settlements.

But sometimes Indians were consulted. As Portuguese India faltered in the seventeenth century before the attacks of the Dutch and English, ready money for the government became increasingly scarce. To deal with this the Indians were represented at least once at a General Meeting (*Junta geral*, or *Conselho geral*) of the town called to discuss new sources of taxation revenue.[60] More often, however, the decisions of such large meetings are signed only by Portuguese.[61] With regard to the municipal council, Indians were sometimes called to give expert evidence, once, for example, on the price of meat, and once on some masonry work, but such participation was not routine and was entirely at the will of the Portuguese on the municipal council.[62] Most of the time, Indians, whether Christian, Hindu, or Muslim, were dependent on informal pressure tactics or individual representation.

The Câmara, on behalf of the Portuguese casados, represented mainly commercial interests. An example is their constant campaign to restrict competition to their own trade. In particular, they got the king to issue a decree meant to stop any converted Jew or Hindu from acting as agent for Portuguese nobles in their trading activities. Similarly, at their instigation the king tried, albeit unsuccessfully, to restrict the trade of the agents of Lisbon capitalists in India, and to reserve to casados the right to hold government tax farming contracts.[63]

The Câmara played a role in other commercial matters also. In 1569, it agreed to a rise of 1 percent in Goa's customs duties, though only because the resulting extra revenue would be used to protect the convoys bringing food to Goa.[64] In the early seventeenth century, the Portuguese empire's revenues fell sharply due to the competition, and hostility, of the Dutch and English. Finally a draconian step, a tax on food, was considered. Free trade in food for Goa had been granted in 1518; it was one of the city's oldest and most prized privileges. Nevertheless, a general meeting in 1623, attended by the viceroy, fidalgos, and the members of the Câmara (no Indians were in attendance, though they presumably also ate food) agreed to this new tax, the colecta.[65]

In 1587 the king made a contract with a group of capitalists in Lisbon, by which, in return for their paying for five years for the fleets sent to India, they were given a monopoly on the indigo trade. This threatened to end the casados' trade in a very lucrative product, for the monopolists could set their own price. They complained violently, and it took the viceroy, the clergy, and some other members of the élite to calm them by assuring them they would get a fair price for their indigo.[66] On occasion the casados even took an initiative, rather than simply reacting to a situation. In the 1590s the casados of Daman, north of modern Bombay, petitioned the king to establish a customhouse in their port. They hoped that if local traders were forced to come there to pay duties the trade of the town would prosper. Years later this was done, but by this time Portuguese power had vanished, and Daman remained small and commercially stagnant.[67]

There had been a threat of violence in Goa during the opposition to the indigo monopoly in 1587, and indeed the casados were ready, if all else failed, to use force to defend their commercial interests. The best example of this was the big fray over customs duties in Cochin, a most interesting dispute, extending over a period of six months, which saw a militant union of Cochin's casados openly challenging the viceroy, while at the same time trying to becloud the issue with proclamations of loyalty to the king and appeals to religious intolerance.

The opposition was over a rise in customs duties for the Portuguese residents of the port. In 1530 D. João III of Portugal had made an agreement with his puppet, the king of Cochin, by which all Portuguese residents of Cochin would pay duties to the king of Cochin. Later the Hindu ruler lowered the rates from 6 per cent to 3½ per cent. This resulted in much cheating and a substantial loss for the Goa customhouse. In 1584, the viceroy, D. Francisco Mascarenhas, acting on the king's orders, made a new agreement with the king of Cochin by which all casados in Cochin were to pay to the king of Cochin a duty of 6 per cent on imports. When the local Portuguese heard of this threat to their trading profits, they met together and agreed to resist by force. They went en masse to see the Portuguese captain of Cochin and complained bitterly that he was going against the concessions confirmed to them by Philip II in 1580 : "that they were very loyal vassals of the king of Portugal, which they had demonstrated many times in his service, when this was possible : that they were not opposing him in this matter : that if he wanted them to pay duties to him they were very ready to agree, but that they refused to pay to a Hindu king like this one". The captain tried to pacify them, but failed. "They all gathered several times in the Câmara and agreed that they would resist by force anyone who tried to give their goods to the king of Cochin", but continued to profess their loyalty to the king of Portugal.

This appeal was both devious and conservative. Rudé has noted how pre-modern crowds often appealed to natural law, or to the king, who would have sided with the crowd if only he had known how his subordinates were oppressing the people. These protestations of loyalty were very much in this tradition. Yet in fact, the viceroy had been acting on behalf of the king and in the spirit of his instructions, so the casados were really opposing the king's wishes. This they tried to disguise by a blatantly discriminatory appeal against paying duties to the Hindu king of Cochin. In fact, they had always paid duties to him. Their real, although unvoiced, complaints were having to pay more duties and other provisions of the decree which would reduce their opportunities for cheating on customs payments.

Their next move was to call a meeting of the native Christians of Cochin and to arm both them and themselves. A total of 15,000 armed men, 10,000 local Christians and 5,000 Portuguese, marched to the church of St. John, put a missal on the altar, and agreed to defend their libertie swith their lives, to persecute those who opposed them, and to stand by their fellows. This extraordinary display of cross-class and cross-racial unity was successful ; the crowd attacked the officials deemed responsible for the new decree, and they fled to refuge in a house of the king of Cochin. In all this the casados were egged on by their women, "who by day and night persuaded their husbands to sustain their ancient liberties". Finally the Portuguese captain, seeing "that general union", suspended all changes in customs duties.[68]

After the rainy season of 1584, the new viceroy, D. Duarte de Menezes, arrived from Portugal to find Cochin in an uproar and the casados still protesting loyalty to the king, "because for him they were all ready to sacrifice their lives and property, but they would give nothing to the king of Cochin". The élite countered indirectly by appealing to God to help them soften the opposition of the casados. To this end they used "sacrifices, prayers, fasts, scourges, and other intercessions". Having impressed the opposition with their fervour the fidalgos then called on them to negotiate. This was agreed on, and twelve men were selected to represent all Portuguese residents in Cochin. These twelve, however, reported back each day to their fellows, all meeting together in the Câmara. Finally both sides gave a little, and in January 1585, peace returned to Cochin.[69]

Neither the casados nor the authorities could afford to be completely inflexible. The casados felt, probably erroneously, that they were forced to resist because they were faced with ruin. But they were careful, first, to try to make their revolt seem both loyal and pro-Christian and, second, not to have to resort to violence. No heads were broken during the affair. The casados were, after all, situated on a hostile coast, and if they were too extreme in their behaviour, the Portuguese authorities would possibly not defend them from *their* proclaimed opponent, the king of Cochin. Nor, of course, could they have hoped to stand up to the king of Portugal's warships and soldiers. On the other side, the authorities had to watch for a possible rapprochement between the king of Cochin and the casados ; the king would have been very annoyed if Cochin were lost. Nor were the élite anxious to use force on their fellow Portuguese. Thus, a compromise was the inevitable result.

A final, and more extreme, example of Portuguese traders looking after own commercial interests was the situation concerning the trade in cloves in the Molucca Islands. This trade was meant to be a royal monopoly, but at least in the 1530s the Portuguese residents there simply ignored this and traded the product on their own. The local officials could do little to stop this. It was dfficiult to get anyone at all to serve or live in the remote Moluccas. Since cloves were virtually the only trade item in those islands, settlers could not be expected to remain there unless they could trade in that item. "For this reason it suited the captains to dissimulate about the offences, and occasionally the injuries, which they received from these men, in order that they would not be left alone in the fort, as indeed, nevertheless, happened many times".[70]

Finally we have to consider several occasions when political or security matters rather than commercial ones were at stake. In these matters the casados were usually not consulted, nor were they concerned with matters which they saw as the prerogatives of their betters. But at times they took a role, either on the instigation of part of the élite, or in a situation of extreme crisis where they felt themselves to be in acute danger, or when they felt their perceived rights were being infringed upon. A good example of the first-two motives acting together occurred in the Moluccas in 1556-57. The local Portuguese captain, an imperious and avaricious man, fell out with the local king of Ternate and had him and his near relatives imprisoned in very poor conditions. Faced with hostility from the locals, the ecclesiastics and the élite in the Portuguese settlement called all the Portuguese residents together and persuaded the captain to moderate the harshness of the king's imprisonment. Despite this concession, forces from the islands of Tidore and Ternate attacked the fort.[71] Fearing that the fort would fall, the local Portuguese united, seized and imprisoned the captain, and, with many apologies, released the king. He forgave them and the war stopped. The Portuguese, however, had now to find themselves a new captain. No fidalgo wanted to accept such a precarious position, but finally one was persuaded to become their temporary leader until the governor in Goa was heard from. Even this was accomplished only after the fidalgo in question had been approached at the door of the church by a procession of all the local Portuguese and clergy, the latter as usual complete with raised crucifix.[72] The deposition of the duly appointed authority was thus soon followed by a craven search for a new authority figure.

The trouble over the governorship of 1527, when two nobles, Pero Mascarenhas and Lopo Vaz de Sampaio, both appeared to have legitimate claims to the office,[73] showed the casados, in this case usually represented by the Câmara of Goa, interfering in high politics and sometimes even without the initiative of the fidalgos. Thus, Lopo Vaz was supported by the officials of the Câmara, and by the principal people (i. e., the fidalgos) of Goa.[74] Later, Pero Mascarenhas made his case in two letters, one to the Câmara and one to all the fidalgos and captains.[75] Later still Mascarenhas got himself released from prison in Cananor after a successful appeal to all the officials and casados of the town.[76] On one occasion during this long dispute Lopo Vaz was opposed by most of the people of Goa, who, after a meeting in the Câmara, went to his house and demanded to be heard. Their spokesman was one of the members of the Câmara, and here that body, rather than the fidalgos, seems to have taken the lead.[77] Soon after, another member of the Câmara addressed Lopo Vaz on behalf of "this city and the people of India".[78] Indeed, the Câmara became very influential in this dispute. Though at one point the captain of Goa, on the Câmara's advice, rejected Lopo Vaz's claims, he later recognized Lopo Vaz as governor after the aldermen agreed to accept him.[79]

This was not the end of the dispute. When both claimants agreed to settle the matter by a special tribunal, which met in Cochin, the casados and the Câmara played an open role. Pero

Mascarenhas had already declared that he considered the Portuguese residents of Cochin to be traitors, for they had not supported his claim. Afraid that he would be chosen and would then avenge himself on them, the Câmara of Cochin lobbied furiously for the selection of Lopo Vaz. All the residents of Cochin "went in barefoot procession to the churches with their wives and children, with many tears and prayers begging God to inspire the judges not to choose Pero Mascarenhas as governor".[80] Lopo Vaz was in fact chosen, and the dispute ended. This sort of open involvement in matters normally reserved to the élite was unprecedented, and should be seen as a consequence of the very unusual situation of Portuguese India in 1527. Normally, the line of succession to the governorship was clearly laid down by royal directives.

In matters affecting themselves however, the casados routinely expected to be heard. As we saw, they had to be consulted before food could be taxed. Similarly, on two occasions the casados of Goa refused to oppose a Bijapuri advance into Portuguese territory. According to them, this sort of martial activity should be done on their behalf by soldiers, not, except in extreme emergency, by them personally.[81] Even in emergencies, they were ready to help with money only on their own terms. In 1586, desperate letters from the captain, bishop, and city of Malacca arrived in Goa. Acheh was besieging this city, the second most important one in Portuguese Asia, and it needed help. The fidalgos of Goa rallied at once, but the viceroy's appeals to the representatives of the Câmara met a cool response. Finally they agreed to call a meeting of the whole casado group and put the matter to them. After a discussion a loan was agreed to, but only because the viceroy offered to secure it with the revenues of the province of Salsette. As a further condition, the casados asked that a fidalgo acceptable to them be given command of the relief expedition. The viceroy had his own uncle in mind for this job, but was forced to bow to the will of the people.[82]

Too much should not be made of these successes by the casados. More often their legitimate complaints were ignored by the élite. The best illustration of this occurred in the 1580s. The Portuguese were involved in a protracted dispute with the Mughal captain of the port of Surat, in Gujarat north of Daman. In 1582 this captain countered the Portuguese fleets by having Daman attacked by land. Considerable destruction resulted, and the casados of Daman frantically appealed for help.[83] Three years later a Portuguese fleet sailed north from Goa to blockade Surat. When it called at Daman for supplies the Câmara protested violently and asked the fleet to turn back, for if it attacked the Mughal captain's ship, the lands of Daman would again suffer the inevitable Mughal reprisal. The captain of the Portuguese fleet was unmoved and replied coldly that he would obey his orders, "and that as for their protests, the viceroy had called together in Goa a council of fidalgos and experienced captains, and he was not going to stay to hear those (craven) things, for he had no choice but to obey; and taking on water and rice he set sail for Surat".[84] Such a rejection of presumptuous casados no doubt met with universal approval from the élite of India.

Many of these cases reveal a considerable lack of patriotism among the casados of Portuguese India, and, indeed, they usually put trade before loyalty. This occurred even in time of war, when one would expect the lines to be more clearly drawn. On the contrary; if, on the one hand, the casados were not consulted about war and peace, on the other, they felt themselves under no obligation to take any notice of the wars of the élite. Reports are legion of Portuguese trading with areas which were at war with the Portuguese, but most demonstrate the insistence of the casados in continuing their very lucrative trade with Gujarat even when Gujarat and Portugal were at war.[85] Similarly, it was not just in the Moluccas that Portuguese casados violated the royal monopoly by trading in spices : reports of Portuguese engaged in trade in pepper and other forbidden products in western India were numerous throughout the sixteenth century, and much of this trade was with areas officially at war with, or at least hostile to, Portuguese India.[86]

One final, and specifically colonial, aspect of the political activities of the sub-élite in Portuguese India needs to be considered. In some areas and at some times the élite were forced to bow to pressure from the economically dominant native population, to the detriment of the Portuguese casados. We have noted that in Goa the native population was usually not consulted, but in Diu, the most important fort north of Goa, the local Gujarati merchant population was so rich and their presence so crucial to the survival of the fort that the élite in Goa usually had no choice but to bow to their demands. This Gujarati power was more negative than positive; thus they played no formal part in the governement of Diu, nor were they represented on celebratory occasions such as the taking of the oath of loyalty to a new king of Portugal.[87] But when they were oppressed, they struck back. The best example of this is the fate of the Câmara of Diu. This representative body, and the title of City for the town, had been bestowed on the Portuguese residents of Diu in 1603, after their repeated requests for this privilege.[88] But the small Portuguese population of about 200, the only ones eligible to vote for the Câmara, misused their position. They raised taxes, abused the Hindu population, and indulged in nepotism. Following complaints from the Gujarati population, the title and the Câmara were removed by the king in 1613.[89]

This study shows the people of Portuguese India acting politically in four distinct ways. One group, the soldiers, was ready to oppose violently their own masters, the captains, when they felt this was necessary for redress. More often, they acted as potential or actual storm troops on behalf of their captains. This first, violent, method was not simply senseless mob anarchy; it had a rationale, for it was the only way to get wrongs righted if a captain was obdurate. The second was a logical extension of the whole recruitment process used by captains in Portuguese India. The other group, the casados, normally was able to make its influence felt through the representative body, the Câmara or municipal council. Only when this failed would they use more open tactics, sometimes led by Câmara officials or by members of the élite. The essential characteristic of such movements is that they were limited to the correction of a specific perceived wrong. They did not aim for greater, or more generalized, political power, nor were they in any other sense liberal, let alone radical. Indeed, their publicity usually appealed to a traditional source.

Did the colonial setting make these activities different from those of similar people in contemporary Europe? It is difficult to find concrete evidence that it did, especially since normally the native population was strictly outside these Portuguese movements. It can only be said that the setting contributed to make Goan society rough, violent, and extravagant, all characteristics of a frontier society, which indeed this was. Specifically, the colonial setting dictated the presence of a large military element in the Portuguese population, while the general setting included the meteorological rhythm which led to large numbers of idle soldiers in the west coast ports during the rains. But, generally, the nature of the Portuguese presence on the Asian coast had two effects on Portuguese society and the political activity of its lower orders. On the one hand it led to considerable Portuguese solidarity in normal social intercourse in peaceful times. The other side of this coin, most manifest in time of stress or danger, was a general touchiness, a readiness to turn to violence, which was at the least influenced by the insecurity of the Portuguese situation in Asia; hence the lack of security in Goa's streets at night and the propensity to avenge an insult with violence. Among the casados this psychic stress is less easy to discern, but symptoms of it can be seen in the brutal treatment of slaves and in the lawless bacchanalian aspects of popular celebrations. This stress became more apparent as the strains increased, notably in the seventeenth century when the English and Dutch began to whittle away at Portuguese possessions in Asia.

NOTES

1. Two of his many books which I have found useful are *The Crowd in History, 1730-1848* (New York, 1964), and *Paris and London in the Eighteenth Century. Studies in Popular Protest* (New York, 1973).
2. See chapter five.
3. François Pyrard de Laval, *The Voyage of François Pyrard de Laval*, 2 vols. (London 1887-90), II, 38; J.H. van Linschoten, *The Voyage of John Huyghen van Linschoten to the East Indies*, 2 vols. (London, 1885), I, 183-84.
4. ANTTDR, XXIV, 34.
5. Ibid., XXII, 97.

6. ANTTCC 1-83-30; ANTTDR, XXXI, 249. On racial divisions in the Portuguese colonial empire, see C.R. Boxer's standard *Race Relations in the Portuguese Colonial Empire, 1415-1825* (Oxford, 1963).
7. Couto, X, 1, 4.
8. C.R. Boxer, *The Portuguese Seaborne Empire, 1415-1825* (London, 1969), pp. 296-97.
9. Linschoten, I, 188. See also pp. 195-96, and the following from p. 193: "The Portingals are commonly served with great gravitie, without any difference between the Gentleman and the common Citizen, townesman or soldier, and in their going, curtesies, and conversations, common in all things".
10. Linschoten, I, 184; Pietro Della Valle, *The Travels of Pietro Della Valle in India*, 2 vols. (London, 1892), I, 157-58. For scenes of people and places in sixteenth century Goa, see the original illustrations in Jan Huyghen van Linschoten, *Itinerario*, 3 vols. ('s-Gravenhage, 1955-57).
11. Detailed descriptions in Couto, VI, iv, 6; Correa, IV, 587: Jacinto Freyre de Andrade, *Vida de Dom João de Castro* (Lisbon, 1822), pp. 247-51, and, based on these, Elaine Sanceau, *Knight of the Renaissance : D. João de Castro* (London, 1949), pp. 186-88.
12. C.R. Boxer, *Portuguese Society in the Tropics : The Municipal Councils of Goa, Macao, Bahia, and Luanda, 1510-1800* (Madison, 1965), p. 40.
13. Pyrard de Laval, II, 215; Jean Mocquet, *Voyages en Afrique. Asie, Indes Orientales and Occidentales* (Paris, 1617), p. 352.
14. LM, XIXB, 607.
15. C.R. Boxer, *The Portuguese Seaborne Empire*, p. 313.
16. LM, XIXB, 607. For a gruesome account of a voyage in 1607-8, see Mocquet, pp. 219-32, and for shipwrecks and disasters see Bernardo Gomes de Brito, *Historia Tragico-Maritima*, ed. Damião Peres, 6 vols. (Porto, 1942-43), or C.R. Boxer's edited translations of parts of this, *The Tragic History of the Sea, 1589-1622* (Cambridge, 1959) and *Further Selections from the Tragic History of the Sea, 1559-1565* (Cambridge, 1968).
17. Boies Penrose, *Goa—Queen of the East* (Lisbon, 1960), pp. 99-101; "Remy", *Goa, Rome of the Orient* (London, 1957), p. 153; Alberto C. Germano da Silva Correira, *La Vieille-Goa* (Bastorá, Goa, 1931), pp. 268, 289.
18. Silva Correira, pp. 274-75.
19. For ill treatment by their superiors see Diogo do Couto, *Dialógo do soldado pratico* (Lisbon, 1937), and Francisco Rodriguês Silviera, *Memórias de um soldado da India*, ed., A. de S.S. Costa Lobo (Lisbon, 1877), pp. 178-90 and *passim*.
20. Pyrard de Laval, II, 131; see also Mocquet, pp. 304-11, 313-17, for the mindlessly brutal conduct of the Portuguese and the mestiços toward their slaves.
21. DI, VIII, 221.

22. Correa, IV, 671.
23. Representative examples of such characterizations are to be found quoted, and corrected, in Rudé, *The Crowd in History*, pp. 7-10, and *Paris and London*, pp. 28-30, 131-32, 293-94, 298.
24. Couto, VI, iii, 5-6.
25. Ibid., X, iii, 4.
26. ANTTDR, XIX, 67-68.
27. Gavetas, VI, 395; for a very jaundiced view of the people, see Barros, IV, ii, 1 : "a gente popular he varia, e inconstante, e amiga de novidades, como pessoas de baixo estado, que sempre o esperam melhorar com a mudança dos tempos. . ."
28. Correa, IV, 671.
29. Mocquet, p. 287.
30. Correa, III, 162.
31. For example, Gavetas, VI, 382, 387-88, 393, 394. For interesting European parallels see Fernand Brandel. *The Mediterranean and the Mediterranean World in the Age of Philip II* (London 1972-3), II, 734-54.
32. Barros, IV, ii, 2.
33. Pyrard de Laval, II, 75, 81, 90, 101.
34. Linschoten, I, 194, 201.
35. Ibid., I, 190.
36. See also ibid., I, 199-201; Pyrard de Laval, II, 269.
37. DI, VIII, 222.
38. For the enquiry, see Gavetas, VI, 370-98.
39. The whole dispute can be traced in Correa, III, 11-25, 79-170, 179-226; Barros, IV, i, 1, 3, 6, 7; IV, ii, 1-6; Couto, IV, i, 1-3, 6, 9; IV, ii, 4-11; IV, iii, 5-9; IV, iv, 1.
40. According to Boxer, *Portuguese Society*, pp. 28-30, they numbered 1,800 in 1540.
41. See Linschoten, I, 228-31, 256; Pyrard de Laval, II, 41, 51.
42. Linschoten, I, 186-88.
43. *Cartas de Affonso de Albuquerque, seguidas de documentos que se elucidam*, 7 vols. (Lisbon, 1884-1935), I, 138-39.
44. ANTTCC, 1-30-36.
45. Ibid.; Linschoten, I, 183-84; DI, II, 548; ANTT; "Colecção de São Lourenço", V, 124v; Barros, III, vii, 9; Couto, V, v, 5; APOCR, V, 193.
46. Pedro Barreto de Rezende, "Descripçoens das Cidades e Fortalezas de India oriental", 2 vols., in Biblioteca da Acadêmia das Ciências, Lisbon, MSS Azuis, 267-68, II, 125-26v.
47. Pyrard de Laval, II, 245-46.
48. Linschoten, I, 266 67. See also HAG, "Termos das Fianças 1626-30", ff. 7-50 for details on Portuguese ships leaving Goa in 1626.
49. For example, Barros, III, vii, 9; "Termos das Fianças", R.J. de Lima Felner, ed., *Subsídios para a história da India portugueza* (Lisbon, 1868), pp. 5-6.

50. Couto, VII, iv, 9.
51. Lima Felner, pp. 8-9; APOCR, IV, 52; III, 403-4, 588; HAG, "Senado de Goa-Cartas Régias", I, 28, 34-34v; HAG, "Livro Morato", ff. 126-27.
52. Couto, IX, chap. 13.
53. Father Felix, "Mughal Farmans, Parwanahs and Sanads issued in favour of the Jesuit Missionaries", *Journal of the Punjab Historical Society* V (1916), p. 8; Artur Viégas, ed., *Relação anual das coisas que fizeram os padres de Companhia de Jesus nas suas missões*, by Fernão Guerreiro, 3 vols. (Lisbon, 1930-42), II, 393-94; P.S S. Pissurlencar, *Agentes da diplomácia portuguesa na India (Hindus, Muçulmans, Judeus e Parses)* (Bastorá, Goa, 1952), p. 99.
54. DI, IX, 289.
55. Pyrard de Laval, II, 246; *The Journal of John Jourdain* (London, 1905), p. 173.
56. Boxer, *Seaborne Empire*, pp. 274-75; see also the following volumes in HAG, for evidence of guild-like organisations in Goa; "Termos das Fianças 1626-30", "Senado de Goa—Termos das Obras, 1654-55", "Senado de Goa—Arrematação das rendas, 1650-1688". Here again, however, it is difficult to tell Portuguese casados from local Christians.
57. On the municipal councils see Boxer, *Portuguese Society*, *passim*.
58. See, respectively, Sylvia L. Thrupp, *The Merchant Class of Medieval London (1300-1500)* (Chicago, 1948), and Violet Barbour, *Capitalism in Amsterdam in the Seventeenth Century* (Baltimore, 1951).
59. See chapter five.
60. Assentos, I, 431-7.
61. For example, ibid., III, 242; HAG, "Livro de ordens régias, 1630-38", ff. 151-52v, and Assentos, 1652-66, ff. 207v-8.
62. HAG, "Acordãos e Assentos da Câmara de Goa, 1592-97", ff. 7v-8v, 162v; ibid., 1597-1603, f. 39.
63. APOCR, III, 753-55; AHU, Codice 282, ff. 16v-16; the following in HAG; "Senado de Goa—Cartas Régias, 1529-1611", ff. 54-54v, 95; ibid., 1594-1603, n.p.; "Alvarás e Provisoes de Sua Magestade, 1617-1652", ff. 73v-74v; "Provisoes a favor da Cristandade", ff. 28v-29v, 49v-50, 85v-86.
64. HAG, "Livro de termos e autos, 1565-72", ff. 182v-83.
65. The following in HAG : "Livro de contrato e concepção da colecta", *passim*; "Alvarás e Provisoes de Sua Magestade, 1593-1781", f. 249; ibid., 1617-52, ff. 28v-29 for D. Manuel's letter of 1518; "Livro de ordens régias, 1630-38", ff. 98v-100.
66. Couto, X, x, 6.
67. APOCR, III, 293-95, 492-63; AHU, Codice 281, f. 139; Artur Basilio de Sá, ed., *Documentação para a história das missões do padroado português do oriente : insulindia*, 5 vols. (Lisbon, 1954-58), V, 218-29,

234; LMBP, III, 155-56; HAG, "Livro de registo das alvarás, cartas etc. de diferentes feitorias, 1667-1684", f. 22.

68. The preceding four paragraphs are based on Couto, X, iv, 13.
69. Ibid., X, vi, 2.
70. Barros, IV, ix, 19.
71. Couto, VII, iv, 7.
72. Ibid., VII, v. 2.
73. See references in note 39.
74. Couto, IV, i, 3.
75. Ibid., IV, ii, 9.
76. Ibid., IV, iii, 5.
77. Correa, III, 186.
78. Ibid., III, 190.
79. Barros, IV, i, 3.
80. Ibid., IV, ii, 6.
81. Ibid., IV. viii, 8-9; Couto, VI, v, 9.
82. Couto, X, viii, 17; cf. ibid., X, x, 4, where the Câmara of Cochin agreed only after a leisurely discussion to help out in Ceylon.
83. Ibid., X, ii, 5-7.
84. Ibid., X, vii, 10.
85. For example, Correa, II, 454-55; Couto, VI, i, 6; ANTT, "Colecção de São Lourenço", III, 53v-54, 227.
86. For example, ANTTCC, 1-25-108; DI, II, 130, AHU, India, caixa 10, letter of February 5, 1633; *Studia, Revista Semestral*, nos. 13-14, p. 80; Gavetas, III, 215.
87. As in Gavetas; III, 88-92.
88. AHU, Codice 282, f. 126; B. Mus., Add. MSS 20863, f. 89v.
89. ANTTDR, XXIV, 23v; AHU, India, caixa 3, no. 4: *Arquivo Português Oriental*, ed., A.B. de Bragança Pereira, 11 vols (Bastorá, Goa, 1936-40), IV, 22, 538-40, 604-5.

THE PORT CITY OF GOA
Policy and Practice in the Sixteenth Century 4

ANY ANALYSIS of Portuguese Goa's role as a port city in Asia in the sixteenth century must inevitably consider Goa's place in the total Portuguese system : Goa was the focus of this system, and most of the profits from it were spent in Goa. There are three categories, the first-two representing official policy, the third practice. First, the Portuguese attempted to take over for themselves, on a monopoly basis, trade in certain Asian products. Second, Asian trade in other products was allowed to continue, but was subjected to Portuguese direction and taxation. Third, at a functioning level both these policies were in fact dependent on the cooperation or acquiescence of various groups of Indians. The three categories obviously overlap and are interconnected; nevertheless the two areas of official policy, and the practical reality underlying both provide useful ways to look at the port city of Goa.

Asian Port Cities

The existing (rather thin) theoretical literature on Asian cities is not very helpful in the case of Goa. The city does not fit in either part of Murphey's division of historical Asian cities, for it obviously was not an inland inward looking city with important political and cultural roles; nor can it be seen as a progenitor of the European-influenced economically-important ports of the nineteenth century, for Murphey finds these to be "situated so as best to serve what had by then [the early twentieth century] become the major areas productive of export goods and the major markets for foreign imports".[1]

In terms of the Redfield and Singer analysis Goa again does not fit. They find four types of city : before the industrial revolution and European colonialism there are administrative cultural cities, and cities of native commerce. Following these there emerge the managerial entrepreneurial metropolis, and cities of modern administration.[2] Goa of course predated the industrial revolution but was a product of colonial expansion, and so makes nonsense of these categories. To the extent that its role was as an administrative centre rather than as an economic focus it belongs if anywhere in the last of these four categories, although its administration was definitely not "modern". Hoselitz usefully points out the difference between parasitic and generative cities, and notes that some early colonial cities contributed to stagnation in their surrounding areas.[3] This comment seems to apply to the more successful Dutch in Indonesia than to the Portuguese in Goa. The Portuguese caused minor restructuring of some Asian trade routes, opened a minor new market for some Asian products, and skimmed a little off the top of some areas of Asian trade. In detail their activities were probably mildly depressive in Malabar, and perhaps a little so in the Deccan, and moderately stimulative in Gujarat : in sum their total impact on Indian trade was very slight. Murphey correctly sees them as "one more in a long succession of traders, manipulating but not transforming".[4] This, however, relates to results, not policy, for the Portuguese did not aim to be like this.

In analytical terms, the most useful way to look at sixteenth century Goa is in the context of the debate over whether colonial port cities were transplants or grafts. Was Goa a city whose functions and characteristics were diffused rather directly from the metropolis, or alternatively should we see it as simply a graft onto an indigenous base? These two notions are obviously ideal types. No port city has ever been purely one or the other, but rather has to a greater or lesser degree combined elements of both. Nevertheless, the distinction is useful, especially if the first category, the transplant, is extended by adding the category of parasites. Without wishing to get too involved in botanical terminology, the crucial distinction is that a transplant moves a system or organism to a new area where it grows relatively independent of its surroundings, its characteristics remaining those it had before it was moved. A parasite is similar in that it feeds on another system but retains unmodified its own characteristics. A graft is generically different, for it encompasses a crucial element of reciprocity; unlike a parasite, it merges and combines with its host. This terminology will be used throughout this chapter.

In the case of Goa all three characteristics can be found. They coincide with the three categories described in the opening paragraph. The Portuguese tried to monopolise trade in some products : this was a policy of transplanting, for the rationale, the method, indeed the very idea of such a system, originated in the metropole. The Portuguese tried to tax and control all other Asian sea trade : this was a parasitic policy, for the Portuguese fed on the activities of others and there was no element of reciprocity. But throughout the century the Portuguese, regardless of what official policy said, were in fact dependent on Asian help and were to a large degree coerced by Asian realities. Thus at a working level Goa as a port city, and the system focused on the city, were grafts. A brief description of pre-Portuguese Goa will provide background for an analysis of Goa and the Portuguese system in the sixteenth century.

Pre-Portuguese Goa

The glamour route in pre-Portuguese Asian trade was that carrying spices from Malacca to Aden, via either Malabar or Gujarat. In this trade the three important Konkan ports of Dabhol, Chaul and Goa played a small role : significantly the wealthy Gujarati ships went direct from Cambay to Malacca.[5] The fifteenth century accounts tend to expatiate on the great ports of Gujarat, and then pass on quickly to Malabar.[6] The difficulty involved in estimating Goa's importance before the Portuguese conquest of 1510 lies in the way in which the earliest Portuguese authors make extravagant claims for fifteenth-century Goa.[7] According to them Goa was the main port of trade for both the Bahmani kingdom and Vijayanagar. As such it imported spices from Malabar, horses from Arabia, and a host of other products; "great trains of oxen loaded with merchandise used to come into Goa from very distant kingdoms in the interior".[8] These claims are apparently influenced by a desire of puff up the importance of the Portuguese conquest of the city. The true position seems to be that none of the Konkan ports in the fifteenth century were very important. The three biggest were Dabhol, Chaul and Goa; given deficient statistics it is pointless to try and rank them.

Politically, militarily and also economically the most important product imported through these ports in the fifteenth century was horses. The Indian subcontinent has never produced good horses. The main local producing area was Kutch, in Gujarat, but horses from there were generally despised. The best came from Persia and Arabia. As Pires said, "The best are the Arabians, next are the Persians and third are those from Cambay. These latter are worth little".[9] The first-two areas however produced very expensive horses. A good one could cost over Rs. 1,000[10] while the average price was around Rs. 600.[11] The import tax on these valuable animals was a major source of revenue in all three ports. In Goa each paid total duties of Rs. 75. At least 1,000 horses came through Goa each year, which meant that about half the total revenue of Rs. 1,80,000 of the city came from this one product.[12]

Yet the main significance of the horse trade in both the fifteenth and sixteenth centuries lay outside of revenues. The Muslim states of the Deccan fought frequent wars with each other and with the Hindu state of Vijayanagar to the south. For these wars good horses were essential, and they could not come overland. Here was the crucial function of the Konkan ports and here was the reason why Goa changed hands so frequently; if the Muslim states (until the 1490s the Bahmani kingdom, then the sultanate of Bijapur) could control Goa as well as Dabhol and Chaul then Vijayanagar must have difficulty in importing horses. The port was conquered from its Hindu rulers by the Bahmanis in 1356. Vijayanagar took over in 1379, and retained it until its reconquest by the Bahmanis in 1471. In the early sixteenth century it was controlled by the Bahmani successor state of Bijapur. The importance of the horse trade was shown in the fact that one of the first things Albuquerque (governor of Portuguese India from 1509 to 1515) had to do after he took Goa in 1510 was to deal with an embassy from Vijayanagar which sought a monopoly on the purchase of horses in Goa.[13]

Goa and Portuguese Policy

From the Portuguese point of view the horse trade with its economic and political advantages was an added bonus rather than the reason for their interest in the port. Indeed, the whole trade of Goa was of little interest. It was location above all else that made Goa attractive, for it was ideally situated to play a crucial role in Portuguese plans. Thus while in modern economic terms Goa's interior, difficult of access and economically underdeveloped anyway, is a crucial disadvantage for the city, in terms of the designs of Albuquerque and his successors this was irrelevant.

This is not the place to repeat a description of Portuguese policies in Asia,[14] yet a skeleton at least must be provided. In the early sixteenth century the main aim of the Portuguese was to take over the spice trade to Europe from the Muslim traders who dominated this product and route. In order to achieve this, the Portuguese claimed the right to inspect all native ships to see if they carried forbidden goods, notably spices but also arms, timber, and hostile foreigners. The great route for spices started in Malacca, the distribution centre for the products of the Spice Islands to the east. From here the ships sailed either to the Malabar coast, itself a pepper exporting area, or to Gujarat, and thence on to the Persian Gulf or the Red Sea.

The first aim of the Portuguese when they reached Asia was to take over this trade for themselves. Such a monopoly would enable them to buy cheap in Asia and sell dear in Europe. The obvious advantages of such a system led the Portuguese to concentrate their efforts toward achieving the monopoly until around the middle of the century. At the same time, however, they tried to direct, tax and control all other Asian sea trade by means of cartazes or passports. Asian traders were forced by Portuguese warships to call at Portuguese forts or ports to pay customs duties and to trade.

To achieve these two aims of monopoly of some trade and control of all other trade certain key points on the Asian coast had to be taken. The crucial ones were Malacca, Aden, Hurmuz and the western coast of India. This was clearly seen by Albuquerque. He took Malacca in 1511 and according to his son,

> often used to say that for the preservation of India, and for the prevention of troubles arising from that territory to the Kings of Portugal, there were four things, of which their possession by the Portuguese must be made very strong and very sure. These were : Aden, in order to have dominion over the Straits of Meca, before the Grand Sultan could forestall them in their seizing it; Ormuz, so as to have supreme rule over the Straits of Baçora; and Diu and Goa, for the sovereignty of all the other districts of India. And with these four places assured to Portugal, and fortified with very strong fortresses, she could avoid many other unnecessary expenses to which she was now subject.[15]

The Portuguese never took Aden, and they had to wait until 1535 to build a fort at Diu. In Malabar they were faced with more or less continual hostility from the Zamorin of Calicut, whose trade they were disrupting (see chapter two). They needed a secure base from which to patrol western Indian waters. Goa in 1510 seemed to be, and indeed was, ideal.

Generally Goa was well located to control the crucial north-west sector of the Arabian Sea. The port was situated almost midway between the key economic areas of Malabar and Gujarat; fleets from Goa could attempt to control the trade of them both. The town was also readily defensible. The area taken by Albuquerque in 1510, and until 1543 the only area controlled by the Portuguese, was an island, in area 48 square miles, called simply Ilhas. The Mandovi river marked the northern, and the Zuari river the southern, coast of the main island of Tissuari, though several smaller islands were also included in the territory of Ilhas. Access to the mainland was available by crossings or fords over the interior creek and the two rivers.

At least by the low standards of the Indian west coast, Goa was a good port. In fact there were two, in the two estuaries of the rivers which surround the island. The harbour in the Zuari estuary to the south was deeper and slightly more sheltered from the southwest monsoon than was the Mandovi estuary to the north, though both were effectively closed between at least June and September by the monsoon.

Goa was protected further inland by the western Ghats, but was not completely cut off. There were a series of breaches in the mountains behind Goa, so that the watershed here was

80 miles inland, rather than the more normal 25 or 35 miles. Similarly, the crest behind Goa is under 2,000 feet, while to the north it is up to 3,000 feet, and to the south over 6,000 feet in places. The effective coastal plain behind Goa stretched inland between 30 to 45 miles. Access beyond this became difficult, despite the comparative lowness of the Ghats here. The valleys tended to be deep gorges two or three miles wide and 1,000 or more feet deep.[16] The main pass used was apparently that defended by the city of Belgaum, 50 miles east-north-east of Ilhas.[17] A contemporary Portuguese described the Ghats as being "so high that it takes two days to get to the top, where there is dense forest".[18] Albuquerque's son described the ascent as being very "steep and rocky," adding picturesquely that "this range hangs over Goa and over the sea like an awning".[19]

The existence of these passes presumably helps to explain Goa's prominence in the fifteenth century, but for the Portuguese they were if anything a disadvantage. They thought they needed the coastal plain for food, but impenetrable mountains behind them would have suited very well. Trade with Bijapur was not essential, but protection from it was. The Ghats were certainly some barrier, yet in 1570, for example, the sultan of Bijapur could get an army of 100,000 men and 2,140 elephants over them.[20]

If Goa was to be a base, it should have adequate supplies of food and water both for its inhabitants and for visiting fleets. Over this matter, as over the importance of Goan trade in the fifteenth century, there is a pronounced difference of opinion between Portuguese writers and foreigners. The Portuguese wax lyrically over the fields and orchards of Goa, its succulent fruit, pure water, and leafy bowers.[21] It seems that these authors are referring to the whole of the Old Conquests, which included the mainland areas of Bardes and Salsette, taken in 1543, as well as the original area of Ilhas. The latter was sourly described by a Dutchman as being "most hillie, barren and wild, and full of waste ground." Nor indeed did Ilhas even have enough good drinking water.[22] Yet even if Bardes and Salsette were more fertile the fact still is, as we shall see, that Goa imported most of its food in the sixteenth century.

There was one other more immediate reason for Albuquerque's conquest of 1510. Muslims in general were disliked by the Portuguese, and the fact that Goa was subordinate to the Muslim sultan of Bijapur was certainly an added incentive. More specifically, the real villains for the Portuguese were "Rumes," European or "white" Turks, and these formidable enemies were powerful in Goa in 1510. Here at least the Crusades still operated; in early 1510 Albuquerque was on his way up the coast to attack his Muslim enemies in Hurmuz. He was diverted when he heard of Goa's large hostile "Rume" population.[23]

This provided the immediate impetus for the conquest, but Albuquerque and his successors soon came to appreciate Goa's other advantages. In sum, Albuquerque took and held Goa because it was an excellent base from which to direct Portuguese activities. The area was relatively defensible, both as an island and as an area backing on the Ghats. Its harbours were adequate, its geopolitical location excellent, and it provided at least some of its own food.

Goa's Maritime Focus

Goa under the Portuguese was always more than itself, more than simply a port city. It was the focus of their entire seaborne empire; its prosperity in the sixteenth century, and its seventeenth century decline, mark fluctuations in Portuguese fortunes in all Asia rather than any change in Goa and its hinterland. It is important to demonstrate how Goa's function, and revenues, came from outside, not from the interior, for this shows clearly its position as a transplant, or parasite, in Asia. We will first demonstrate generally how Goa focused outward,[21] and then attempt to disaggregate this data to show how some Portuguese policies were transplants and some parasites.

The clearest indication that we have to show which way Goa was focused in the sixteenth century comes from figures for customs duties collected on inland and sea trade. The result demonstrates clearly the predominance of the latter. There were a total of eight passes from Ilhas to the mainland: Panjim, Naroa, the Dry Pass ("passo seco," also called São Bras), Benastarim (also called Santiago), Ribandar, Daugin, Carabolim (also called

São João Bautista) and Agaçaim (also called São Lourenço).[25] At each of these passes a capitation tax was collected by the Portuguese in command there, and customs duties were levied on goods. We have spotty data for the sixteenth and early seventeenth centuries, which in sum shows relative volumes of inland and sea trade. But several qualifications need to be stressed. Some items, such as food, gold, and horses were exempted from paying duties. Some of the available figures have been rounded off in the sources. The rates charged are not always completely certain. We have no indication of the extent of smuggling. We do not have breakdowns for receipts at each of the eight passes. Sometimes we have only a gross figure for the "passos," and sometimes only four or five are listed. This is because sometimes two passes were combined, while others perhaps did not always have customs houses.

In the later years of the sixteenth century the customs house of Goa was usually rented for around Rs. 3,24,000 a year, and the passes for Rs. 7,200.[26] With duties at the customs house of 4½ per cent, we have a total value of goods reaching and leaving Goa by sea of Rs. 72,00,000. The duties at the passes were 1½ per cent, so goods passing through them were worth Rs. 4,80,000.

In 1605 the customs house was let for Rs. 3,45,450.[27] Around 1600 the passes were let for about Rs 6,750.[28] Thus, with duties now of 6½ per cent, we have sea trade valued Rs. 53,14,000 and land trade at Rs. 4,50,000. In 1616-17 the figures for sea trade are Rs. 36,00,000 and for land trade Rs. 6,00,000.[29]

What these figures show is the clear dominance of sea trade. It will be noted that the ratio between sea and land trade goes from about 15 : 1 toward the end of the sixteenth century to 12 : 1 early in the seventeenth century and to 6 : 1 in 1616-17. The relative imbalance remains large throughout, and is simply huge in the sixteenth century, the period with which we are concerned. What the figures also show, even with due allowances for our weak data, is a fall in sea trade, and a rise in land trade. This indeed is what one would expect, given Dutch and British competition from the early seventeenth century.

As another demonstration of the focus of trade in sixteenth century Goa, we have figures for the total values of all routes sailed from Goa around 1600. The total figure differs from those just given based on customs duties because this one includes goods and routes which did not have to pay duties. The total of all routes was around Rs. 97,50,000.[30] As opposed to this, we can put land trade at a maximum figure of about Rs. 7,50,000, and the ratio at 13 : 1.[31]

A second run of statistics again shows Goa's orientation to the sea and the profits the Portuguese made from coercively focusing western Indian trade on Goa. These figures are for the total revenue of Goa, and for customs duties as a percentage of this total. They show that in 1545 63 per cent of Goa's revenue came from customs duties. In 1586-87 the figure was 57 per cent, around 1600 56 per cent, and in 1616-17 44 per cent.[32] A point to be noted is that Goa in fact got less of its total revenue from customs duties than did other Portuguese areas. Figures for the whole Portuguese empire in 1586-87, and around 1600, show over 65 per cent of the revenue of the whole empire coming from customs duties.[33]

A further indication of Goa's outward focus is seen in the way the Portuguese were even prepared to watch with equanimity while Goa's production of food, and trade with the interior, declined. Easy pickings from the functioning of the whole system made these things seem to be irrelevant. The much praised lushness of the lands of the Old Conquests should have meant that Goa would be able to feed itself, at least after 1543 when Bardes and Salsette were taken. The area of Ilhas included 31 villages and in the sixteenth century these paid a total revenue of about Rs. 15,000.[34] Salsette had about 50 villages, and Bardes about 40, and in sum these paid revenue of about Rs. 97,500.[35] This seems to indicate considerable agricultural productivity in these two new areas, as is borne out also by the lyrical descriptions in sixteenth century accounts, and by on-site inspection today. Yet increasingly Goa was dependent on food imports.

The method the Portuguese used to make sure local traders came to Goa was to group them in convoys of several hundred local merchant ships, and then sail them under guard of perhaps ten Portuguese ships to Goa. This method protected local traders from the endemic piracy of the area, and reduced the probability of local traders going to other ports. We have records of food convoys coming from Kanara, south of Goa, from the early 1560s.[36] In the 1570s and 1580s three or even four food convoys came each year from Kanara to Goa.[37] This points to very large food imports from Kanara, yet in 1570 the well-informed chronicler Couto claimed that most of Goa's food came overland from the area of Bijapur between Goa and the Ghats. He described how "the bulk of the food for this city comes from the lands of Bijapur whence come everyday like trails of ants vessels loaded with rice, wheat, grains" and various vegetables.[38] There are also accounts of food coming to Goa from Portuguese possessions in southern Gujarat.[39] In the 1630s the value of the rice trade from Kanara alone was put at Rs. 4,50,000.[40] With rice worth around Rs. 12 a candil (500 Portuguese pounds) this makes a total import of 1,80,00,000 regular pounds from Kanara for a total population of the Old Conquests of about 270,000 people.[41] Thus we have 66 pounds (30 kgs) per head per year being imported from this one area alone. The average consumption in India today is 200 kgs per head per year. Given that Kanara apparently supplied only a fairly minor part of total food imports, it is clear that Goa imported most of its food.

There were infrequent complaints about this situation[42] but by and large the Portuguese were prepared to live with it. The seas were relatively secure for Portuguese convoys in the sixteenth century, and profits from other sources meant Goa could afford to import food. In any case, the Portuguese had not come to India to be farmers. In the palmy days of the sixteenth century large areas of what good land there was in Ilhas were apparently used for pleasure gardens and most Portuguese preferred to live by trade, not from agriculture.[43]

A similar outward focus can be seen more generally in the virtual end of trade inland with Bijapur, and instead a strong emphasis on policies designed to direct sea trade to Goa. The city did not graft on to existing patterns, but rather attempted, by transplanting a metropolitan idea, to change Goa's whole focus. As we noted, Goa's importance as an entrepot in the fifteenth century is difficult to assess. However, it is clear that, regardless of volume, it was a true entrepot, channelling exports from the Bahmani kingdom and Vijayanagar, and importing goods for these interior markets. This is proved by two short descriptions of Goa's pre-Portuguese trade. One noted large imports into Goa, so that "large caravans of merchants came from the interior country in quest of it, and brought other commodities in exchange".[44] Another commentator claimed that "Cargoes of rice are taken from here (Goa), and great trains of oxen loaded with merchandise used to come in to Goa from very distant kingdoms in the interior".[45] Given the problem of crossing the Ghats, these accounts are probably exaggerated, yet clearly Goa had a considerable inland trade in the fifteenth century.

Under the Portuguese this changed. Immediately after the conquest Albuquerque sent out sea patrols to force ships into Goa, especially those carrying horses. His aim was to use the horses as a bait to tempt the trade caravans to come again from Bijapur and Vijayanagar.[46] This attempt failed, partly because of Bijapuri hostility, but more because the Portuguese soon found that they did not need inland trade, the only exception being the horse trade. Thus later sixteenth century accounts describe Goa's "livings and daylie traffiques" to Bengal, Siam, Malacca, Cambay, China and every way "both North and South"[47] but not east to Bijapur. Only food seems to have come from there. Even the local Kanarese and Deccani shop keepers in Goa sold goods from Gujarat, China, and Bengal but not from the interior.[48] An early seventeenth-century author put the products of Bijapur in their place when he listed Goa's imports from the mainland as wheat, meat, horse food, and "much more important than all this, rowers for the oared boars".[49] The Portuguese were always scornful of Bijapuri products, preferring to buy cotton cloths, the main trade item, from Gujarat.[50]

Transplants and Parasites

Having demonstrated Goa's general orientation, we must now investigate which aspects of Portuguese policy were intended to achieve a monopoly of a particular trade, and which simply aimed to control the trade of others. In terms of the botanical analogy, the former can be classified as transplants, the latter as parasitic. The distinction between the two lies in the fact that a transplant, an attempt to take over trade in a product, actually involved the Portuguese themselves : in its purest form, the spice trade, *they* bought the product and transported it to Europe. Control over someone else's trade is different, for here the Portuguese simply battened on to an existing system : here they were merely parasites. The two categories are related however, for they both were based on European ideas of monopoly, and owed nothing to Asian realities. At least in theory, they were purely ethnocentric, as can be seen from this paraphrase of the semi-official Portuguese justification for them. According to Barros, the Portuguese were in Asia lords of the sea, and made all other ships take a safe-conduct or cartaz from them. Ships trading to enemies of Portugal could be seized on sight. By common law the seas were open to all, but this applied only in Europe to Christians, who were governed essentially by the principles of Roman Law. Hindus and Muslims, on the contrary, were outside Roman Law as they were outside the law of Jesus Christ, which all men must keep to avoid the eternal fire. Further, Hindus and Muslims had no claim to right of passage in Asian waters, because before the arrival of the Portuguese no one claimed the seas as hereditary or conquered property. There being no preceding title, there was no present or future right of passage.[51]

Transplants

The classic examples of attempted transplants, efforts to monopolise a product, were the spice and horse trades. As we noted, the horse trade from the Persian Gulf had been important in the fifteenth century. Once Albuquerque had taken Goa, the Portuguese took pains to foster this lucrative trade. Any ship bringing in to Goa at least ten horses from Hurmuz paid no customs duties on its remaining cargo, and merchants who left Goa having bought five horses were given concessions in the customs houses covering Goa's land trade.[52]

The Portuguese also tried compulsorily to centralize the whole horse trade in Goa, so that any ship coming from South Arabia or Hurmuz with horses for the Indian market could come only to Goa. In practice, this was never enforced very strictly, despite some success in the early years. The main chink in the system was that in places outside Goa local rulers and local Portuguese wanted horses for themselves, and both these groups had occasionally to be conciliated by the Goa government for political reasons.[53] Another difficulty was that the religious authorities disapproved of Christians selling horses to infidels, as this was forbidden by a Papal Bull. Finally God and Mammon were both served by a Provision of 1574. Portuguese were no longer allowed to take horses inland from Goa to sell in Bijapur. Instead the Bijapuri merchants would have to come to Goa to buy them. Thus the consciences of the Portuguese were freed from the sin of taking horses to Muslims.[54]

From this trade the Portuguese crown derived both financial and political benefits. Horses paid no import duties in Goa, but when they left paid about Rs. 75 each. As early as 1513-14 this trade was worth Rs 9,900 to the Goa customs. By 1523 this had risen to Rs. 36,000 and in the 1540s to Rs. 1,26,000. Later in the century receipts fell sharply to around Rs 30,000 as more and more horses entered India at ports other than Goa.[55] An undated account of about 1580 says that Goa made Rs 18,000 from the horse trade, but Chaul, another Portuguese port, made Rs 28,500.[56]

The political benefits, although less direct, were equally substantial. Vijayanagar and the five Deccan sultanates were frequently at war with one another in the sixteenth century before Vijayanagar's great defeat near Talikota in 1565. Each was prepared to offer large concessions to the Portuguese in return for a denial of horse sales to their enemies. In 1547 Bijapur and Vijayanagar were at war, and both sides offered the Portuguese ample and cheap supplies of provisions in return for a monopoly of the horse trade. The Vijayanagar offer was more attractive, and they received the desired assurance that the Portuguese would sell no horses to Bijapur.[57]

Important as the horse trade was, it was the attempt to control the spice trade which engaged the Portuguese most, especially in the first-half of the sixteenth century. It is here that the Portuguese themselves were most involved; this therefore is the purest example of a transplant. The subject has been investigated rather fully in several recent works.[58] Generally it is clear that this attempt to enforce a monopoly had only spotty success. For most of the sixteenth century more spices got to Europe outside the Portuguese system than were carried by the Portuguese around the Cape of Good Hope. But we are here concerned with Portuguese policy, not its success, and this policy was one of pure monopoly.

Another major part of Portuguese policy should also be included in this transplant category. The government laid down that on particular routes only a specifically licensed ship could sail in a particular year. Cargo space on these monopolistic voyages thus sold at a premium. In the earlier part of the century these voyages were done in royal ships, but from the 1540s the Crown took a more passive role and simply licensed individual Portuguese to undertake them. Sometimes voyages were given as dowries for noble but impoverished girls. Some official posts included several voyages as part of their salary. Thus the captain of Malacca had voyages to seventeen different places.[59] But usually these voyages were sold to the highest bidder, and from this again the Crown made large profits. In 1602 a voyage from Goa to Macao and Japan raised over Rs 52,500 for the Crown.[60]

Parasitic Policies

Revenues from the horse and spice trades were rather substantial, and indeed when combined with the customs duties discussed earlier they made up most of Goa's state income. However, the Portuguese also profited from a related but different kind of activity, one which was purely parasitic, in that they simply skimmed off money without themselves providing any reciprocal advantages, or engaging in the trade themselves. This was the well-known cartaz system.[61] Some idea of the parasitic profits the Portuguese made from this system can be gained by noting the money the Portuguese received from outlying forts and ports. These profits were derived mostly from customs duties and so were based on their trade control policies. Portuguese armadas in the sixteenth century attempted to force all local traders to call in at Portuguese forts and pay duties. When this system worked, especially at Hurmuz and Diu, it worked extremely well and produced large surpluses which were sent to Goa for general state needs. Again, therefore, we see Goa as the centre of Portugal's seaborne empire, getting revenue not from its own economic activity so much as from its political and military dominance in Asian waters.

Albuquerque took Hurmuz in 1515, but left a puppet local king in formal control. In the 1520s around Rs 45,000 came to Goa from Hurmuz, for the local rulers gave the Portuguese Rs 90,000 a year, and only about half of this was needed for local Portuguese expenses.[62] In 1543 the Portuguese took over control of the Hurmuz customs house. Portuguese expenses at this fort were now around Rs 39,000 a year, but the average proceeds from the customs house for the years 1543-50 were Rs 1,35,000. The Portuguese thus had a profit, based on political and military control, of over Rs 90,000 a year in the 1540s.[63]

The fortress of Diu, at which the wealthy Gujarati ships bound for the Red Sea had to call, produced similar profits for Goa and the Portuguese empire. Portuguese expenses here ran at around Rs 75,000 a year in the second-half of the sixteenth century, but as early as 1556 revenues were around Rs 1,32,000. In the 1580s two different accounts put Diu's revenue at over Rs 1,95,000.[64] Diu thus produced a profit of Rs 90,000 to Rs 1,20,000 in a good year.

Portuguese Policy in the Sixteenth Century

This unique system of extortion, of an empire based on control of and profits from local trade done by others, was never fundamentally questioned by the Portuguese in the sixteenth century. It was, however, modified during the century, and at times the modus operandi were discussed.

The Portuguese began the century in a very militant hard-nosed fashion, burning Muslim ships on sight, massacring their crews, and generally running riot in Asian waters. Most of these efforts were directed to trying to obtain a monopoly over the spice trade. But there was simply too much ocean to patrol effectively, and too many determined and skilful Asian traders. From the 1560s Portuguese control over the spice trade weakened as increasing amounts ran the gauntlet and reached the

Red Sea safely. In response the Portuguese empire became more autonomous, their policy less concerned with the trade to Europe and more concerned to increase Asian trade than to worry over much about where it went. In terms of the analysis, the attempt to control the spice trade was a pure transplant ; later in the century Portuguese policy became more parasitic, yet in neither case was there the element of reciprocity and synthesis which would make these policies of graft. As examples of changes later in the century, the prohibition on trade to the Red Sea (in goods other than spices) was dropped, and by and large Portugal aimed to foster local trade, and private Portuguese trade, so that customs revenues would increase. The system softened a little, but of course retained its basically coercive character ; after all, Gujarati ships only put in to Diu because they had to, not because they wanted to trade there.[65]

The coercive element was central to the Portuguese effort from the very earliest days. Their policies, especially in the early days, were affected by virulent anti-Muslim feeling, and with Muslim traders dominating many of the best routes the Portuguese for both economic and religio-political reasons saw no alternative but to drive out the hated "moors." Later there were debates about how best to operate the system. In particular, many writers claimed the Portuguese had acquired too many forts. More attention should be paid to the armadas, while the large number of unprofitable forts could be abandoned, retaining only the centrally important ones of Goa, Diu, Hurmuz, Malacca, and perhaps all or part of Ceylon.[66] But (a point we will return to shortly) no one suggested a radically new policy of peaceful trade on a basis of equality, a policy of graft, as was attempted by the British and the Dutch in the seventeenth century. For the Portuguese, the Muslim traders were hated too strongly, and the easy profits of the mid-century years were too great, for any fundamental change to occur. The existing policies, despite some perceived weaknesses, had produced the lavish life of profligate Golden Goa. The fullest extent of projected changes recommended only mild reforms to make the coercion more efficient.

The comparative success of the sixteenth century was based on sea power, and this meant the Portuguese were seldom tempted to try large land conquests. Again we see clearly the Portuguese evaluation of their aims. An extensive land trade from the interior of Goa would have required at least some extension of Portuguese control inland. But as the system emerged of reliance on sea power and trade, any idea of internal expansion ceased; it was not needed in economic terms, and in any case political relations with Bijapur made any attempt at expansion both unnecessary and hazardous. The Portuguese were happy enough to sit in Goa. Any attempt at expansion would only affect Goa's food supplies.[67] Goa's strong position in the crucial horse trade, and the defensibility of her territories, especially Ilhas, discouraged Bijapur from mounting large attacks on the Portuguese; this tacit modus vivendi suited both sides. Once the initial conquest had been accepted by Bijapur, the break came in the 1570 siege by Bijapur, and this seems to have been a response to recent Portuguese thrusts initiated by an impetuous new governor. Peace was soon restored, and was never seriously threatened again.

Grafts

Beneath the level of official Portuguese policy there was another, one of practice, and here we can see Goa continuing on as an Asian city. Portuguese policy attempted and in a few cases succeeded in making a radical break with fifteenth century conditions, but at an immediate level Portuguese Goa remained a graft on the western Indian coast. The Portuguese lacked the means to put into effect fully their policies. At an individual, functioning, level there were innumerable compromises and accommodations with Indian realities. As these have been considered rather fully elsewhere,[68] a brief summary designed to point the contrast with official policy, and thus show how Goa was on different levels transplant, graft and parasite, is all that is needed here.

In very many areas the Portuguese were dependent on Asian manpower and expertise. Most of their pepper exports came from Malabar; this pepper was grown and transported to the coast by Indians. Goa's food arrived in Indian ships, and came from an Indian-ruled area. Gujarati merchants provided loans to the Portuguese state in the parlous days of the early seventeenth century. The vast bulk of Goa's official revenues were collected by Hindu tax farmers. The whole Portuguese system, ostensibly so coercive and ethnocentric, based entirely on ideas from the metropole, was in fact dependent on acquiescence and cooperation of various groups of Indians.

Socially and religiously also there was more reciprocity and grafting than unidirectional impact. The Portuguese have been widely condemned for their forced conversions to Christianity. Leaving aside the vexed question of the role of force in these conversions, the important point is that the curious hybrid which is Goan Christianity owes much to the Hinduism which it supplanted. As only one example, caste divisions continue after several centuries to be important in the lives of Goan Christians.[69] The inexorable rhythm of Goa's monsoon climate modified Portuguese social life, as contemporary accounts make clear. Food, dress, architecture, the daily round of life all were as much Indian as Portuguese, producing a graft which was neither completely one nor the other. Indeed, in the seventeenth century Portuguese policy veered towards acquiescence in or even encouragement of this fact, most notably in an official proposal to force rich young Hindu girls to marry poor Portuguese, thus literally grafting political control with economic power.

In the context of this divergence between policy and practice it is tempting to speculate generally about Portuguese aims. We have shown that in practice the element of reciprocity, of graft, underlay the policies of transplant and parasite. But what if policy and practice had been convergent, if in fact a *policy* of graft had been followed, which led to the Portuguese trading in Asia on a basis of equality and reciprocity? Despite tacit changes in policy later in the sixteenth century this was never tried, but could it have worked? Obviously no definite answer can be given, but it may be suggested that the Portuguese had certain advantages which *could* have led to their being able to compete successfully with other Asian sea traders. As a state enterprise they probably had the upper hand in terms of manpower, capital, and coherent policy making, over their less centralised rivals. Their ships were apparently more technologically advanced than those of the competition, and certainly they were better armed. At the least, the Portuguese should have been able to do very well in the trade to Europe, given the cost-effective benefits of the all-sea Cape of Good Hope route as compared with the route through Egypt.

It is true, no doubt, that those who in 1500 dominated international trade in Asia, especially the various Muslims from Malabar, Gujarat and the Middle East, would have competed strongly to avoid losing their position. However, it is worth remembering that in nearly every case it was the Portuguese who fired first. Had they not started the violence, the possibility is that Muslims would have competed hard, but peacefully, and indeed would no doubt have found the presence of the Portuguese creating some commercial advantages as well as debits.[70]

A comparison with the achievements of the Dutch and British in India in the seventeenth century would be instructive here : overall, I feel such a comparison would bear out my tentative claim that cooperation, trading freely like any other group of foreign merchants, might have given better results. But given all we have said about the ethnocentric baggage which the Portuguese brought with them to Asia, such a policy was never likely to be tried.

The Portuguese Failure

What can be said with certitude is that in fact Portuguese policy was something of a failure, despite the political and military advantages they enjoyed. On at least three economic criteria Goa was inferior to the great Mughal port of Surat in Gujarat. First, Surat boasted a powerful and integrated merchant community which regulated merchant and even artisan affairs, and safeguarded commercial interests when they were threatened by the state. There was nothing in Goa comparable with Surat's mahajans. All we can find are artisan level organizations, and some hints of cooperation amongst various Indian commercial groups, notably the Saraswat Brahmins and the Gujarati vanias. None of these acted as important checks on official rapacity, nor apparently did much to regulate merchant affairs.[71]

Second, the trade of Surat was based overwhelmingly on local products, notably cotton cloths, and also saltpetre, indigo dye, and numerous handicrafts. The rest of Surat's exports came from northern India. Her imports were few—some copper, spices, luxuries—and consisted mostly of bullion. Compare this sort of generative trade with Goa's, which had few links with its hinterland, and was based on compulsion.

A third area of comparison is that of total values of goods paying customs passing through Goa and Surat. The best figure we have for Goa's trade at its height is about Rs. 67,50,000. For Surat the only figure we have is Rs. 1,00,00,000 for the year 1644.[72] These two figures tend to bear out Surat's economic dominance. Furthermore, Goa's figure is one for a uniquely central focus of Portuguese activities in all Asia, and few of its trade goods arrived there freely. On the other hand, any one trading to Surat did so of his own free will, while Surat was also only one, though admittedly the biggest, of a number of important ports in the gulf of Cambay. Thus despite Portugal's massive military efforts, her local port still had much less trade than did Surat.

The ephemeral character of Goa's prosperity, and its total dependence on trade directed by Portuguese fleets, was starkly revealed in the seventeenth century. Faced with Dutch and British competition, open attacks, and regular Dutch blockades of Goa itself, the port declined precipitously. This can be seen even in the figures through to 1616-17 quoted earlier. Between 1600 and 1616-17 Goa's sea trade fell from Rs. 53,14,000 to Rs. 36,00,000. In 1635 the total value had fallen further, to Rs. 27,60,000.[73] By the mid 1680s Goa had sunk to a very third rate position, with a total value of trade of only Rs. 10,20,000.[74]

More concretely, in 1638 the Portuguese wanted to send the king of Acheh a diamond encrusted cross, which would cost Rs. 15,000, "but we don't have any state revenues here for this expense as everyone knows how exhausted the royal treasury is. And in this city there is no merchant or jeweller who will make the cross on his own account, with repayment to be made from the reciprocal presents of the king of Acheh".[75] Meanwhile in Surat the great Jain merchant Virji Vorah died around 1677 worth about Rs. 75,00,000.[76] Around 1700 the total trade of Surat was worth over Rs. 1,60,00,000 in a good year.[77] The official Portuguese attempt to transplant a monopoly of some trade items to Asia, and to profit parasitically from the trade of others, was revealed as a failure. The subsequent history of Goa, though less extreme and profligate, shows official policy becoming more realistic. In practice even in the sixteenth century there had been compromises with Asian facts. Increasingly in the seventeenth century Portuguese policy accepted this, and Goa became officially, as well as de facto, a graft on to the parent Indian stem. The reconquest of 1961 was merely the culmination of this process.

NOTES

1. Rhoads Murphey, "Traditionalism and Colonialism : Changing Urban Roles in Asia *Journal of Asian Studies* XXIX (1969), *passim*, and especially pp. 67-73, 75.
2. Robert Redfield and Milton Singer, "The Cultural Role of Cities," *Economic Development and Cultural Change* III (1954), pp. 56-59.
3. Bert Hoselitz, "Generative and Parasitic Cities," *Economic Development and Cultural Change* III (1955), pp. 279-80.
4. Murphey, p. 82.
5. Tomé Pires, *The Suma Oriental of Tomé Pires*, 2 vols. (London, 1944), II, 269.
6. For example, R.H. Major, ed., *India in the Fifteenth Century* (London, 1857), *passim*; *Travels of Ludovico de Varthema* (London), 1858), pp. 115-6.
7. For example, *The Book of Duarte Barbosa*, 2 vols. (London, 1918-21), I, 175; Pires, I, 58; Barros, II, v, 2; Castanheda, III, viii; *The Commentaries of the Great Afonso Dalboquerque*, 4 vols. (London, 1875-84), II, 95.
8. Pires, I, 57.
9. Pires, I, 21.
10. Ibid.
11. Barbosa, I, 94.
12. Barros, II, v, 2; Castanheda, II, xvi.
13. Castanheda, III, cxviii; Affonso de Albuquerque, *Cartas*, 7 vols. (Lisbon, 1884-1935), II, 341-3; Correa, II, 376, 379.
14. On this topic see three sound articles in the *New Cambridge Modern History* : I.A. MacGregor, "Europe and the East," II, 591-614; J.B. Harrison, "Asia and Africa," III, 532-58; J.B. Harrison, "Europe and Asia," IV, 644-71. See also my *Merchants and Rulers in Gujarat : The Response to the Portuguese in the Sixteenth Century* (Berkeley, 1976), chapters I and II.
15. Albuquerque, *Commentaries*, IV, 24.
16. O.H.K. Spate, *India and Pakistan : A General and Regional Geography* (London, 1967, 3rd ed.), pp. 665-7.
17. Couto, V, ix, 8.

18. Castanheda, III, xv.
19. Albuquerque, *Commentaries*, 95.
20. J.N. da Fonseca, *An Historical and Archeological Sketch o the City of Goa* (Bombay, 1878), p. 149.
21. Albuquerque, *Commentaries*, III, 278, 288; Albuquerque, *Cartas*, I, 21-3; Orlando Ribeiro, "Originalidade de Goa," *III Coloquio Internatio-nal de Estudos Luso-Brasileiros. Actas* (Lisbon, 1957, vol. I), p. 171; Anthony D'Costa *The Christianization of the Goa Islands* (Bombay, 1965), p. 18; João Manuel Pacheco de Figueiredo, "Goa Pré-Portuguesa," *Studia* 12, 13-14 (1963, 1964), no. 12, pp. 145-51.
22. *The Voyage of John Huyghen van Linschoten to the East Indies*, 2 vols. (London, 1885—), I, 182-3; *The Voyage of François Pyrard of Laval*, 2 vols. (London, 1888), II, 7 and f.n., 28, 70-71.
23. Barros, II, iv, 5; II, v, 4; Barbosa, I, 175-7; Pires, I, 56.
24. Reference must be made here to a recent article by A.R. Disney which runs counter to the theme of this chapter : "The Portuguese Empire in Indla, c. 1550-1650 : Some Suggestions for a Less Seaborne, more Landbound Approach to its Socio-Economic History," in *Indo-Portuguese History : Sources and Problems*, Delhi, 1980. He finds an increasing land orientation in the Portuguese Indian empire, especially in the Old Conquests and the North (Daman and Bassein). This is no doubt true in the seventeenth century, as indeed my own figures later in this chapter show, and is to be explained by the fact that the Portuguese really had no choice once the Dutch and British took over most of their sea trade. But I doubt that this land orientation began before the seventeenth century, and thus I feel the argument of this present chapter can stand, especially as it refers specifically to Goa and not to the North.
25. Albuquerque, *Commentaries*, III, 280-5. This is a citation to an appendix in this book which prints extracts from an important manuscript of 1635 : Pedro Barreto de Rezende, "Descripçoens das Cidades e Fortalezas da India Oriental." F.P. Mendes da Luz, ed., "Livro das cidades e fortalezas que a Coroa de Portugal tem nas partes da India . . ." photographic facsimile in *Studia* VI (1960), f.5r.
26. J.M. do Carmo Nazareth in *O Oriente Português* VI (1909), p. 17.
27. ANTTCC, 3-27-44.
28. Luis de Figueiredo Falcão, *Livro em que se contém toda a Fazenda e real patrimonio dos Reinos de Portugal, India e Ilhas adjacentes* (Lisbon, 1859), pp. 75-78.
29. AHU, India, caixa 6, account of 19 Nov., 1618.
30. Pedro Barreto de Rezende, "Descripçoens das Cidades e Fortalezas da India Oriental," 2 vols in Biblioteca da Acadèmia das Ciências, Lisbon, II, ff. 125-6r.
31. Using figures above based on customs duties, and roughly doubling them to include goods like food which were not taxed at this time.

32. See *Merchants and Rulers in Gujarat*, p. 32 and f.n. 4.
33. Ibid.
34. B.H. Baden-Powell, "The Villages of Goa in the Early Sixteenth Century," *Journal of the Royal Asiatic Society* (1900), p. 263 f.n.; Barros, II, v,; AHU, codice 500, ff. 1-2r; ibid., India, caixa 6, account of 19 Nov., 1618.
35. See the last two sources cited in f.n. 34.
36. Couto, VII, x, 2; VII, x, 19; Viceroy Noronha to king, 30 Dec., 1564, in *Studia* III (1961), p. 77.
37. Couto, IX, caps. 18, 19; X, iii, 9.
38. Couto, VIII, cap. 34.
39. LM, XIXB, 470.
40. Barreto de Rezende, II f. 126r.
41. Teotonio R. de Souza, "Glimpses of Hindu Dominance of Goan Economy in the 17th Century," *Indica* XII (1975), pp. 29-30.
42. City to king, 1602, "Senado Régias," II, n.p , HAG; Couto, VIII, cap. 34.
43. François Pyrard, II, 28.
44. Albuquerque, *Commentaries*, II, 95.
45. Pires, I, 57; see also Barros, II, v, 2.
46. Albuquerque, *Commentaries*, III, 39-40.
47. Linschoten, I, 184.
48. Linschoten, I, 256-7; Albuquerque, *Commentaries*, III, 293-302 (see note in f.n. 25).
49. Ibid., III, 304.
50. Ibid.; AHU, India caixa 3, no. 116; António Bocarro, *Decada 13 da História de India* (Lisbon, 1876), p. 336. See also chapter five for the trade to Gujarat.
51. Barros, I, vi, 1.
52. "Provisoes, Alvarás e Regimentos," I, f, 22, HAG; Correa, II, 74. For a good general discussion of the horse trade see C.R. de Silva, "Empire, Horses and Trade," International Conference on Indian Ocean Studies, Perth, 1979, Proceedings of Section III.
53. F. Paes, "Tombo de Ilha de Goa. . .," *Boletim Vasco da Gama* no. 62; "Livro de Leis de D. Manuel," ff. 137-7r, ANTT; P.S.S. Pissurlencar, ed., *Regimentos das Fortalezas da India* (Bastorá, Goa, 1951), pp. 271, 314-5, 383; APOCR, III, 14; ibid., III, supplement, pp. 7, 711-2, 493; AHU, codice 281, ff. 264r, 359-60; R.J. de Lima Felner, ed., *Subsídios para a história da India Portugueza* (Lisbon, 1868), p. 116.
54. Couto, VIII, caps. 29, 33; IX cap. 17.
55. Marechal Gomes da Costa, *Descobrimentos e Conqüistas*, 3 vols. (Lisbon, 1927-30), III, 306; ANTTCC, 1-30-36; Gavetas, V, 247; AHU, codice 500, ff. 1.2r; Figueiredo Falcão, p. 75; DUP, I, 231. From this last it appears that only about 300 of the 1500 horses

which came to India each year entered through Goa in the late sixteenth century.

56. B. Mus., Addl. 28433, f. 75r.
57. Correa, IV, 601, 604-5, 621; R.J. de Lima Felner, pp. 255-57. See also the treaty of this same year with Vijayanagar's vassal the raja of Kanara, in Couto VI, v, 5. For similar rivalry between Bijapur and Vijayanagar and offers of concessions in 1513-14 see Correa, II, 376, 379; Castanheda, III, cxviii; *Alguns Documentos do Archivo Nacional da Torre de Tombo* (Lisbon, 1882), pp. 367-9, 381; Albuquerque, *Cartas*, IV, 341-3.
58. See footnote 37 of chapter two for references.
59. See *Merchants and Rulers in Gujarat*, pp. 36-7, and f.n. 18.
60. "Provisoes dos Vicereis," I, f. 96, HAG.
61. See *Merchants and Rulers*, pp. 30-56.
62. ANTTCC, 1-30-36.
63. Simão Botelho's tombo in Lima Felner, pp. 87-92.
64. ANTTCC, 1-102-47; B. Mus., Addl. 28433, ff. 76, 78, 116r; AHU, codice 500, f. 98; ibid., India, caixa 6, report of 17 Jan. 1619.
65. I.A. MacGregor, pp. 610-11; J.B. Harrison, III, 536-9; *Merchants and Rulers*, p. 51.
66. *Merchants and Rulers*, pp. 53-6.
67. For example, Couto, V, ix, 10; Pyrard, II, 132-3.
68. See chapters three and four in this book, and T.R. de Souza, "Glimpses of Hindu Dominance of Goan Economy in the 17th Century," *Indica* XII (1975), pp. 27-35; G.V. Scammell, "Indigenous Assistance in the Establishment of Portuguese Power in Asia in the Sixteenth Century," *Modern Asian Studies* XIV (1980), pp. 1-11.
69. See, for example, B.G. D'Souza, *Goan Society in Transition* (Bombay, 1975).
70. Niels Steensgaard has a useful passage dealing with this matter in his *The Asian Trade Revolution of the Seventeenth Century* (Chicago, 1974), pp. 83-85, and his whole book relates directly to the comparison with the Dutch and English mentioned in the following paragraph. See also C.R Boxer, *The Portuguese Seaborne Empire* (London, 1969), pp. 37-38. J.H. Parry has recently argued the contrary case. He says the Portuguese had to use force (though he does not condone this) because of the advantages the Muslims enjoyed, and their opposition to the Portuguese. See *The Discovery of the Sea* (London, 1975), pp. 214-6.
71. See works cited in footnote 68 above.
72. Ali Muhammad Khan, *Mirat-i Ahmadi* (Baroda, 1965, trans. M.F. Lokhandwalla), p. 193.
73. Barreto de Rezende, II, f. 150.
74. LM. LII, f. 338.
75. Fazenda, V, 30r.

76. M. Escaliot, "The Plunder of Surat by Shivaji in 1664," *Indian Antiquary* VIII (1879), p. 260.
77. Ashin Das Gupta, "Gujarati Merchants and the Red Sea Trade, c. 1700-1725," in Blair B. Kling and M.N. Pearson, eds., *The Age of Partnership : Europeans in Asia before Dominion* (Honolulu, 1979), p. 124.

BANYAS AND BRAHMINS
Their Role in the Portuguese Indian Economy

5

Uncle Edward Ponderevo "became in those days the terror of eminent historians. 'Don't want your drum and trumpet history—no fear,' he used to say. 'Don't want to know who was who's mistress, and why so-and-so devastated such a province : that's bound to be all lies and upsydown anyhow. Not my affair. Nobody's affair now. Chaps who did it didn't clearly know . . . . What I want to know is, in the Middle Ages, Did they Do Anything for Housemaid's Knee? What did they put in their hot baths after jousting, and was the Black Prince—you know the Black Prince—was he enamelled or painted, or what? I think myself black-leaded—very likely—like pipe-clay—but *did* they use blacking so early?' "

(H.G. Wells, *Tono-Bungay*.)

HISTORIANS of South Asia probably have a rather unfavourable image of the Portuguese in India, to the extent that they have any image at all of them. Vague recollections of da Gama's high-handed approaches to the Zamorin of Calicut trigger memories of his barbarous treatment of Muslim merchants and traders at sea, and so of the Portuguese attempt, in the sixteenth century, to monopolise the spice trade and exploit other Asian traders. A.K. Priolkar has reminded us of the activities of the Inquisition in Goa, and again this brings on memories, of forced conversions and religious discrimination. In the works of Hunter, Whiteway, and Danvers we read of the profligacy of life in Golden Goa, though this perhaps does not prompt quite such an unfavourable response. The picture here is fleshed out with hazy recollections, of Portuguese housewives prostituting their numerous female slaves and themselves drugging their husbands the more easily to disport themselves with their lovers. And all these half-formed images are filtered through a general liberal dislike of twentieth century Portuguese colonial policies : of their stubborn retention of their

Indian colonies, and of their exploitation of the African "Overseas Provinces."

My aim here is not to refute these images : all too often they are borne out by the facts. The Portuguese military-political-religious presence in sixteenth century India was in many respects an exploitative, racist, distasteful presence. Most of the horror stories are true. Da Gama *did* massacre, often with considerable barbarity, boatloads of Muslim merchants and hajjis. In 1524 Indian traders who had tried to evade the Portuguese monopoly of the pepper trade *were* stoned to death in the streets of Goa. Non-Christians, and Christians considered to be schismatic *were* at the least discriminated against. The Portuguese kings *did* try to enforce their grandiloquent claim to be "Lord of the Conquest, Navigation and Commerce of Ethiopia, Arabia, Persia and India." In 1546 Governor de Castro *did* proudly tell the king how he had sent "D. Manuel de Lima with twenty foists to cover all the gulf [of Cambay] and burn and destroy the whole coast, in which he very well showed his diligence and gallantry, because he caused more destruction and damage on the coast than was ever done before, or ever dreamt of, destroying every place from Daman up to Broach, so that there was no memory left of them [sic], and he butchered everyone he captured without showing mercy to a living thing. He burnt twenty large ships and 150 small ones . . . and the town squares were covered with bodies, which caused great astonishment and fear in all Gujarat".[1]

But there was more to it than this. Historians of British India have increasingly come to stress the survival or florescence of various groups of Indians beneath the Imperial umbrella; readers of this literature will not be surprised to find that Hindus played a dominant role in the Portuguese Indian economy. Indeed, given Portugal's home population of about one-half million, and the huge area in which she operated in the East, stretching from the Cape of Good Hope to Japan, it is clear that she had no choice but to co-operate with local inhabitants and allow them a considerable share at least in the economic sphere. But this fundamental aspect of the Portuguese Empire has been virtually ignored. Most of its historians have preferred to concentrate on the activities of the Portuguese themselves, and to describe the empire solely from this viewpoint.

In seventeenth-century India, at least in the economic sphere, there was opportunity for all, and while there was religious and racial intolerance this did not prevent non-Portuguese from playing a dominant economic role, and of profiting from the extra opportunities made available by the Portuguese.

In theory non-Christians were discriminated against by various laws. These laws were designed to encourage conversion to Christianity; by denying non-Christians the right to practise their religions freely, by excluding them from the top level of the political process, and by attempting to limit their economic activities, it was hoped that conversion would seem desirable. In the religious sphere attempts were made to prohibit Hindu ceremonies in Portuguese territory, while Hindu orphans were liable to be educated and baptized by the Jesuits.[2] Politically, non-Christians, and indeed most non-Portuguese, were excluded from the highest level of decision making. In the seventeenth century in Goa a General Council was occasionally called to decide some important matter affecting all Goa's inhabitants—an example was that called before the *colecta*, a tax on food, was instituted in 1623. At such councils as many as one hundred people might attend, including the Governor and the officials, the *fidalgos* or "nobles", representatives of the clergy, both of the Orders and the seculars, and the members of the Municipal Council. But I have found only one meeting when representatives of the Hindu community were present.[3] In the economic sphere also non-Christians were subject to discrimination. In 1595 they were officially forbidden to trade to any ports except those on the Western Indian coast. No Portuguese official or captain was allowed to employ a Hindu as his agent in his trading ventures, or as his assistant. Nor were they allowed to hold government contracts—but as we shall see this was never observed. It is notable that such measures were often prompted or supported by the Goa Municipal Council. This body consisted of lower-class Portuguese, and these were the people who competed, though unsuccessfully, with the Hindus in business and trade.[4]

Despite these, and many other, laws, it can be shown that Hindus held an important place in the Goan, and indeed the Portuguese Indian, economy, and that this economic power had political implications. By investigating the roles of two particular Indian groups in Portuguese India it can be shown that beneath the hard-nosed, unidirectional upper level of Portuguese policy there existed a lower level, where Indians and Portuguese worked together and in which the Indians in fact often held the upper hand. We will restrict ourselves primarily to the economic roles of Gujarati *vanias* ("banyas") and Saraswat Brahmins, as these are the two groups for which comparatively more data is available : some mention will also be made of other Hindu groups in Goa, and of Kanarese and Malabari merchants and traders south of Goa.

In the city of Goa, taken by Albuquerque in 1510 and the capital of the Portuguese Asian empire from 1530, it is clear that the dominant native élite was the Saraswat Brahmins.[5] This position in fact apparently continued until the end of Portuguese rule in 1961. Sea trade from Goa was controlled by Portuguese and vanias, but in the city and lands of Goa the Saraswats held the upper hand. Thus in the 1580s the owners of the spice shops and groceries of Goa "are commonlie the Bramenes, which serve likewise for Priests and Idolatrous Ministers, and have their shops throughout the Cittie. In every place and corner, and under pentises, whereby every man may have to serve him at his need."[6] These brahmins were also the controllers of most of the land of Old Goa, in Ilhas, Salcette, and Bardes. Further evidence of the wealth of the brahmins, and others, comes from a curious proposal in the 1640s. The municipal council of Goa wrote the king of Portugal suggesting that he pass decrees that "no brahmin or kshatriya or member of any other caste who is rich or has property"[6] might marry his daughter to anyone except to a Portuguese born in Portugal : such people must also bequeath all their property to these daughters. The advantages of thus coopting the wealthy Hindu population would be numerous. The Portuguese population would receive an economic shot in the arm from the dowries and legacies their Indian wives would bring with them. Their wives and children would of course be Christian, and thus the true believers in Goa would grow in numbers. And this measure would result in the lands of Goa being populated by the earnestly desired "white men". The sons of such marriages would be prepared to bear arms, "which the brahmins do not do because it is not their custom".[7]

By far the best evidence concerning the economic position of the native population in the Goan economy, and especially the dominance of the brahmins, comes from the relatively copious data on who collected taxes. Most sources of government revenue in Portuguese India were farmed out by the Crown to the highest bidder, and were called collectively *rendas*.[8] We have good data on these for the period 1600-1670. The usual term was three years. The holder had to have a guarantor, and deposit a surety before taking up his contract. If no satisfactory bid was received the renda in question was run by Crown officials —as frequently happened during our period with the Goa and Cochin customs duties. Other rendas included the tax on food shops, on cloth sales, on tobacco sales in Goa, Daman and Bassein, on official paper, on opium sales, on spice sales, inland customs duties, the special tax on seed pearls, and many other sources.

During an average year between 1600 and 1670 rendas to the value of Rs. 5,16,600 were let out. This figure on its own means virtually nothing. How large a sum was this in terms of the total economy of Portuguese India? Given the scarcity of statistics available, and the dubious nature of some of those which we do have, it is impossible to say precisely; we can only present some other data and so try to give some impression of the comparative size of revenue from rendas in the whole economy. In terms of the total revenue of the state, these revenues were very important. For the whole of Portuguese India, from East Africa to Malacca, the government's receipts were : around 1600, Rs. 17,77,800; in 1635, Rs. 17,77,900; around 1680, Rs. 10,97,742. The total listed expenses of the whole State in these three years was : Rs. 11,78,390; Rs. 16,73,180; Rs. 6,81,995.[9] To give other figures : in 1654 a large brand new galleon of 1,000 tons was worth Rs. 44,100; the average price of a candil of rice (about 500 lbs avoirdupois) in the first-half of the century was a little under Rs. 15; the average price paid by the Portuguese for their pepper between 1611 and 1626 was Rs. 85 a candil; the usual annual customs revenue of the great Mughal port of Surat in the 1640s was Rs. 2,50,000.[10] These figures are sufficiently random, but they show that the average annual value of all the rendas, Rs. 3,16,600, was a considerable sum for the day, and that any group holding a large number of them must have possessed considerable capital, and occupied an important place in the total Portuguese Indian economy.

During the period 1600-1670 there were ideally 782 holders of rendas, for there were 34 different ones, and each would be let 23 times. (In fact the actual total was much less, as several rendas vanished; either this particular tax was stopped, or more likely its administration was taken over by Crown officials.) We know the names of 355 of these holders, or 45.4 per cent of the ideal total—a remarkably large sample for seventeenth-century India. Of these 355 holders, 284, or 80 per cent, were Hindus, and 71, or 20 per cent, were Christians. Even more interesting, 178 were held by people who can be identified from their names as being Saraswat Brahmins. The caste or group of the other 106 Hindus is less easy to determine; but 16 were Gujarati vanias; 11 were kayasthas; three were Marathas; and 17, called Chatim, were either vaisyas or goldsmiths—if the latter, they claim to be brahmins, but are considered to be sudras by others. Saraswats held 62.7 per cent of all the rendas held by Hindus, and this one community held exactly half to the whole total. There is no case of a Muslim holding a renda. Of the total number of rendas, Saraswat Brahmins held 50 per cent, other Hindus 30 per cent and all Christians, both Portuguese and local converts, 20 per cent.

The dominance of the Hindus is modified only a little when we consider values. We know the value of 327 of the rendas let during our period, or 41.8 per cent of the ideal total. The total value of these 327 was Rs. 49,68,106; rendas worth Rs. 17,28,900 were held by Christians, and Rs. 32,39,206 by Hindus. The percentages are respectively 34.8 per cent and 65.2 per cent. Within the Hindu total, 69 per cent by value were held by Saraswats : in terms of the total value of all the rendas the Saraswats held 44.9 per cent, the Christians

34.8 per cent and other Hindus 20.3 per cent. It is thus to be expected that the average value of a renda held by a Christian was higher than that held by a Hindu, and also that the average for Saraswats was higher than that for other Hindus. Such is indeed the case. The average value for Christians is Rs. 26,593, for all Hindus Rs. 12,357, for Saraswats Rs. 13,525, and for Hindus other than Saraswats Rs. 10,390. Within this last group, the average value for Gujarati vanias is Rs. 13,003; for kayasthas Rs. 8,352; for Marathas Rs. 5,967; and for people called Chatim Rs. 16,320. The average value of all the 327 rendas for which we have details is Rs. 15,192. There seems to be an explanation for the high value of the Christian rendas, which we will present in the next paragraph, but the important point is not so much the average value as the total. Here we see clearly how dominant Hindus were in this important part of the Portuguese Indian economy. By number they held four-fifths of the rendas, by value nearly two-thirds.

The Christian holders were divided between Portuguese and local converts, but usually we have no way of telling which were which—João de Souza, for example, could be either. But it seems that a small number of very large rendas were held by Portuguese businessmen or their agents. This was especially so in the case of the Goa and Cochin customs duties, which were let out as one renda, and were in 1605-08 worth Rs. 3,45,495 a year. This renda was twice held by B.S. Correa, who was of Portuguese Jewish origin,[11] and another holder was Dr. Cosmo Gonsalves Freire, also clearly Portuguese. A small number of such valuable rendas inflates the whole average figure for Christians. Here it is significant that the median figure for the value of a renda held by a Christian is Rs. 6,900, 26 per cent of the average figure of Rs. 26,593. (For Saraswats the median is Rs. 5,250, 39 per cent of their average, for other Hindus Rs. 3,300, 32 per cent of their average.) The lowness of the median as compared to the average points to a small number of large values distorting the average for the whole group. Further, it can be shown that Indian converts held much less valuable rendas. Nineteen of the Christian holders are identified after their names as being brahmins ('bragmane' or 'bramane') or gancars, village heads;

thus we have "João Bernandes, bragmane e gancar". Such people were local converts. The average value of rendas held by these 19 people was only Rs. 11,278—less than the average for all Hindus or for Saraswats, and only about two-fifths of the average for all Christians. There is, in fact, a rough progression of values. The most valuable were held by Portuguese. The large number of middle value were held by Saraswats, and the least valuable by other Hindus and local Christians.

One other piece of evidence may be presented which again shows the dominance of Hindus, but in a reverse way. We have details on 105 contracts, let out by the Municipal Council between 1654 and 1664. There were very minor contracts, for small taxes, and repairs to buildings. The average value of the 57 on which we have information is only Rs. 755. Of the total of 105 where we know the names of the holders, 72 or 68.5 per cent were let to Christians, all almost certainly local converts, and only 33, or 31.5 per cent, to Hindus. Only one of these was a Saraswat Brahmin. It is true that the Municipal Council members tended to be rather bigoted, and so perhaps they favoured their co-religionists, but they needed the money, and it is most likely that they let their contracts to the best bidder, be he a Christian or a Hindu. It seems that most local Christians lacked the resources to be able to bid for one of the larger contracts, but could bid, and successfully, for these smaller ones.

We have very little information on who held rendas at times other than 1600-1670, but such as we have seems to show that the situation in these seventy years was not atypical of the first-two centuries of Portuguese rule in India. We have the names of 23 holders of rendas in 1528, and 1541-43. Of these, seven were Christians, seven were other Hindus, eight were Saraswats, and one was Muslim. For 1702-05 we have 37 holders : six were Christians, six were other Hindus, and no less than 25 were Saraswats. These figures are so few and scattered that they show little except that Hindus held the majority of rendas throughout these two centuries, and that possibly the Saraswats moved from a position of equality with Christians and with other Hindus in the first-half of the sixteenth century to equality with these two combined in the first-seventy years of the next century, and to almost complete dominance by the end of this century. But this interpretation of "the rise of the Saraswats" awaits further research for confirmation. Over the seventy years for which we have detailed information the three groups remained rather lumpishly in the same relative positions throughout, with no significant changes in their respective shares being evident.

One would expect to find the dominant Hindu groups, and especially the Saraswats, organized in bodies, the aim being to increase the share of the rendas held by a particular community, and protect the interests of those members of the community who held rendas. Unfortunately there is no direct evidence of the existence of caste panchayats in seventeenth-century Goa, but it would be extra ordinary if they did not exist. And there is evidence of various holders of rendas, and their relatives and caste associates, operating in a way rather analogous to the modern system of interlocking company directorships. When a person bid successfully for a renda, he was obliged to nominate one or several guarantors, and these were often relatives, and often also holders of other rendas. The following example illustrates this. (I have put the spelling in the seventeenth century documents in brackets after the modern spellings of some of these people's names.) In 1640 Narayana Naik (Narna Naik) and Dhaktya Naik (Dacutea Naiq), who were cousins, held the rendas for spices, food shops in Ilhas, and duties on gold. Among their guarantors were another Narayana Naik, son of Narsu Naik, who formerly held the tobacco renda, and a third Narayana Naik, son of Raulu Naik, who used to hold the renda for the tax on silk and cotton cloths. In 1641 our second Narayana Naik, son of Narsu Naik, again got the tobacco renda, and among his guarantors were our third Narayana Naik, son of Raulu Naik, plus an uncle of this, our second Narayana Naik, and thirdly Mangoji Sinai Vaglo (Mangogi Sinay Vagallo), son of Mangogi Sinai and a gancar of the village of Cortalim in Salcette. This last-named twice in the 1640s held the renda for silk and cotton cloths.

There are glimpses of other, more institutionalized, groups existing in Portuguese India. The Gujarati vanias in Diu, Ormuz and Goa had their own "captains", who acted as agents for all Gujarati merchants in dealing with the Portuguese authorities in these places. These captains occupied recognized, honoured positions and were treated with considerable respect by the Portuguese.[12] At the other end of the economic scale from the opulent vanias there is evidence of guild-like organizations among artisans and handicraftmen in Goa. A number of people, often both Hindu and Christian, would bind themselves to be collectively responsible for each other. To quote a typical case, in 1644-71 goldsmiths of the Rua de São Paulo agreed to be jointly responsible for all gold, silver, jewels and pearls entrusted to any of them for use in their work. Thus if one absconded with some gold and jewels which had been given to him to be made up into jewellery, the other seventy people would have to compensate the owner. We may note that only one of these goldsmiths was Christian—the rest were Hindu, most of them called, as one would expect, Chatim.[13]

Given the discriminatory legislation outlined at the beginning of this chapter, the dominance of Hindus in this important part of the Portuguese Indian economy requires some explanation, especially as these rendas were under direct Crown control—if discrimination was to work anywhere one would expect it to work here. The answer lies partly in the sheer numerical dominance of the Hindus, at least early in the seventeenth century, and more importantly in their control of capital. In 1625 it was estimated that there were about 60,000 native Christians in Goa, and over 100,000 Hindus. In Goa town in 1635, and excluding convents, there were said to be 3,500 hearths, 800 of which belonged to Portuguese householders. And all the people living in one of the 2,700 non-Portuguese hearths would be non-Portuguese, while most of the members of one of the 800 Portuguese hearths would also be non-Portuguese.[14] In the other areas of Western India where we have details for rendas, Chaul, Daman, Bassein and Diu, the Portuguese population was even smaller. More important in our explanation, most of these Portuguese were officials, soldiers, or clerics. The number of businessmen, who ordinarily would be the only ones capable of holding rendas, was small. Most of the Portuguese participated in trading ventures, but the holding of a renda would involve the owner in considerable supervisory work and constant attendance in Goa or the other cities. Among the Portuguese only the businessmen were able to fulfil these conditions. As to the local converts to Christianity, they seem to have come mostly from lower economic or caste groups in Goa, and do not appear to have improved their position much after their conversion. We noted above that the very minor contracts let by the Municipal Council were held mostly by Christians; they seem to have been unable to compete with Hindus for the most valuable rendas. Those local Christians who did hold rendas usually held only the less valuable ones. Many of these were converted brahmins and the fact that they had themselves officially described as "bramanes" points to their feeling more affinity with their ex-caste members than with their fellow Christians of lower caste origin. Albuquerque noted in 1514 that "these native Christians are poor people" ("eses christaos da terra pouca fazenda tem"), and, except for the few high-caste converts, their status did not improve noticeably over the next two centuries. The Muslims, according to Albuquerque, were the only rich people in Goa at the time of the conquest, except for the vanias.[15] A general massacre of Muslim men occurred after the Portuguese took Goa for the second time in 1510, and they seem never to have returned to Goa in large numbers. References to Muslims resident in Goa are almost non-existent, and in Diu they were well outnumbered by the vanias. The reader will remember that we found only one case, out of a total of 521, of a Muslim holding a renda, and this was in the 1540s. One other piece of evidence points to the same lack of economic importance of the Muslims. We have a detailed list of the purchasers in Goa of the cargoes of two Gujarati ships seized by the Portuguese in 1630. Christian buyers numbered 401, Hindu 378, and Muslim 10. Christians bought goods worth over Rs. 30,000, Hindus over Rs. 25,500, and Muslims Rs. 180.[16] Generally, the Crown officials had no choice but to let off rendas to Hindus; the Hindus were faced with weak or no competition for them.

Among the local inhabitants of Goa the Saraswat Brahmins were clearly the most wealthy, but they appear to have been outranked within the town by a smaller but wealthier group of outsiders, the vanias ("banyas") of Gujarat. These were members of a group of castes, both Hindu and Jain, who have been important in trade and commerce in Gujarat for centuries. A report of 1646 estimated that there were 30,000 vanias in the whole of Portuguese India, primarily in Goa, Diu, Bassein and Daman. Some were reputed to have capitals of Rs. 4,00,000 at their disposal.[17] In Goa, Gujaratis held rendas, and in the town there was a street of their shops, selling mostly cloths and precious stones. Gujaratis lent money to the Goa government, sold goods to it, and owned ships in partnership with Portuguese.[18] In one case in 1659 four of them were asked to raise perhaps Rs. 1,50,000 in a very short time. They did so, and it is significant that in this crisis it was the vanias who were called on not the Saraswats or the Portuguese businessmen.

The heart of the Gujarati economic position in Portuguese India was their sea trade, based on the products of their homeland. As a viceroy once told the king, the Gujaratis "have always been the most lucrative group of all in terms of the customs revenues".[19] Gujarati ships sailing from the Gulf of Cambay to the Red Sea or elsewhere were forced to call at Diu, at the mouth of this gulf, to pay duties to the Portuguese. After all expenses had been met at Diu, the surplus, around Rs. 1,00,000, was sent to Goa to be used in general state expenses.[20] As the total Portuguese revenue from the Goa area around 1600 was about Rs. 6,10,000, the profits from Diu were most welcome.[21] Portuguese revenue from the puppet state of Hurmuz, at the mouth of the Persian Gulf, was also dependent on Gujarati trade. By an agreement of 1523 the king of Hurmuz was to pay his Portuguese masters Rs. 90,000 a year. But most of the king's revenue came from trade with Gujarat, and his taxes on this trade, so it was also provided that if this trade was blocked, the king had to pay the Portuguese only Rs. 37,000.[22]

Of equal importance to the Portuguese was the central role of Gujarati products in Portuguese Asian trade. As they well knew, "the source of all the trade of India begins with these cloths and goods of Gujarat".[23] Of all the routes sailed from Goa around 1600, those to Gujarat ranked above all others, including even that to Portugal. The annual capital involved on Goa-Gujarat trade was well over Rs. 40,00,000 a year.[24]

Most of this trade was with the great port of Cambay, whence came two or three convoys of merchant ships each year. The goods involved were Gujarati, but the return cargoes were, to the distress of the mercantalist-influenced Portuguese, largely bullion. The owners were also apparently mostly Gujarati, but many of the inhabitants of Goa, both Portuguese and Indian, invested in these fleets too.[25] From Goa some of these goods were distributed in Asia, but the bulk of them from the Cambay convoys were sold in Goa to Portuguese, and by them sent on to Portugal. Thus we have a detailed inventory of the privately owned part of the cargoes of three Portuguese ships bound for Lisbon in 1630. There were relatively trifling amounts of spices and drugs, and a host of odds and ends, but Gujarati cloths were overwhelmingly dominant. Of these the three ships carried a total of nearly 4,500,000 yards.[26]

On a more microcosmic level, there is abundant evidence of individual Portuguese and Gujaratis co-operating for mutual profit. These "unofficial" contacts were based on the importance of Gujarat as a money market. To sixteenth century European observers Gujarat was a fabulously wealthy country. An Italian in the 1560s said flatly that "In fine, the kingdom of Cambai (Cambay=Gujarat) is a place of great trade, and hath much doing and traffique with all men. . ." At the same time a Portuguese claimed that "if in any land it can be said that gold and silver flow, it is Cambay". A Jesuit observer of 1600 described how Gujarat, then and for centuries past, received streams of gold from the Red Sea, the Persian Gulf, and China.[27] In 1537 as in 1646 Portuguese writers wondered at the wealth of the vanias, some of whom had capital at their disposal of Rs. 2,00,000, and at the later date much more.[28] On a wider level, we may simply note that Gujarati sea trade at the end of the sixteenth century was worth about

Rs. 8,00,00,000 a year. By comparison the average capital on English East India Company of London ships between 1601 and 1640 was about Rs. 2,00,000.[29]

Private Portuguese were more than willing to utilize Gujarati capital. There were several hundred Portuguese resident in the town of Cambay, at the head of the Gulf of Cambay, living by trade. Many others visited the town regularly, while still others traded there from Goa. All these Portuguese used Gujarati brokers and their capital in Cambay. There seem to have been more Portuguese traders resident in Cambay than in any other non-Portuguese Asian city, so we can assume that a considerable proportion of private Portuguese trade in Asia was financed with Gujarati capital.[30]

Other contacts, and instances of co-operation, were legion. On one occasion a request to the king by a Portuguese official for a reward for his past meritorious services was supported by two petitions from Diu, one from the Christian residents and one from the Hindus and Muslims.[31] Portuguese officials employed Gujaratis as their agents, even though this was an illegal practice.[32] Even less legal was collusion between Portuguese and vanias to cheat on customs payments. In Cochin, Portuguese paid lower duties than did Indians. What could be more natural than for a Portuguese to sell his goods to a vania at a slightly inflated price, and in return pass these goods through customs as belonging to a Portuguese?[33]

What were the political implications of the economic importance of these two, and other, Hindu groups? Hindus were consulted, although only very infrequently, by Portuguese governmental bodies. Thus a large meeting in 1632 was convened to consider how to raise more government revenue, and to alleviate the famine of that year. The Indian groups present voted against removing the existing tax on rice imports, presumably because one of the proposed substitutes for this tax was a fee to be levied on non-Christian marriages.[34] More informal influence was presumably used at times, although it goes unrecorded. But the clearest instance, demonstrating Portuguese reliance on Indian co-operation or at least acquiescence, occurred in the 1560s. A new viceroy, D. Francisco Coutinho, described to the king how he found Goa much depopulated when he arrived, with "the villages ruined, the fields flooded, the river clogged up, and the Hindu inhabitants absent". The reason was a decree of the last viceroy, D. Constantino de Braganza, who had ordered that all Hindus who had left Goa to avoid forcible conversion must return or their property would be forfeit. The rate of exodus had increased after this. D. Francisco, with the consent of the archbishop and the "educated men", repealed the offending decree and promised to restore all confiscated property if those absent returned within six months.[35] In fact, Portuguese Goa could not survive without its Hindu population.

As we noted in chapter four, the city was also dependent on Indians for its food. Then as now Goa was a rice-deficit area. Most of the rice came from Kanara, south of Goa. Two or three convoys of at least a hundred small coastal craft made the voyage each year, protected by Portuguese warships.[36] These fleets of Indian-owned and manned ships were literally Goa's lifeline. Even in time of war between the states of Portugal and Kanara the Portuguese authorities at Goa had no choice but to allow this trade to continue. In 1655 the two were at war, but the Kanarese merchants either did not know or did not care. They continued to come to Goa with cargoes of rice. The Portuguese protested that such conduct was "much to be abominated" and "very damaging to the credit and reputation of the arms of his Majesty, who God protect", but bought the rice anyway.[37]

This dependence was also in evidence in the organization of the pepper trade. Along with all other spices, trade in pepper was meant to be a monopoly of the crown of Portugal. Recent research has shown how ineffective the Portuguese in fact were in stopping the spice trade to the Red Sea, let alone that within Asia.[38] Less attention has been paid to the way in which the Portuguese were dependent on various sorts of Indians for the pepper cargoes for their own ships. Thus in the Kanara production area the local rulers in the seventeenth century tended to be difficult. They were far from being in awe of the Portuguese, and their independent and even insolent attitude annoyed the Portuguese intensely, apart from increasing the price the latter had to pay for pepper.[39]

The pepper from this area was taken to Goa, and from there exported to Europe. This Kanara-Goa trade was, for a period of at least thirty years, handled by one Hindu family. This was a large business : in 1617 the Portuguese treasury owed them Rs. 1,80,000, Rs. 1,72,500 as payment for a pepper contract, and Rs. 7,500 for a loan which the family had given to help outfit a Portuguese armada. I have been unable to identify the caste of this family, although we have the names of three of them : Babaji, Govinda and Pondhia. Their importance to the Portuguese was shown when the king, Philip II of Portugal, pointed out, "it is necessary that both these particular Hindus, and the others living in Goa and other towns of Portuguese India who are beneficial to my revenues, be favoured so that they will be encouraged to keep serving me. . . ."[40]

The group best able to turn economic power into political influence were the Gujaratis. Indeed, their political influence was even greater than might be expected, because the prime economic role of the Gujaratis lay in sea trade, and the revenue of the Portuguese state was, as we have noted previously, heavily dependent on profits from, and especially taxes on, this sea trade. Thus customs revenues were crucial, but some other items of revenue were also based on sea trade, especially the sales of the rights of trade to certain areas, which often yielded large sums. In 1602 a voyage from Goa to Macao and Japan raised about Rs. 55,000 for the Crown.[41] There were also occasional windfalls from sea trade, especially when a ship trading outside the Portuguese system was captured. Such prizes could be worth Rs. 1,00,000 or more for the king, apart from the bounties for the crews of the Portuguese ships involved.[42] A further basic point is that this customs payment came from inter-Asian trade, not the trade with Portugal, which was mostly tax free, and which declined during the sixteenth century.[43] We have seen the position of Gujarati goods and merchants in this Asian trade; the political consequences are obvious, and will only be touched on here.

War between Gujarat and Goa was a disaster for the Portuguese state and its merchants. The war of 1613-15 cost them both dearly, for the vital convoys from Cambay to Goa were stopped, and it was only by desperate measures that a cargo for the homeward bound ships could be procured.

Normally the private part of these cargoes consisted mainly, as we have seen, of cloths and various other products from Gujarat, and the substitute cargo of 1615, inferior cloths from the Deccan, was far from satisfactory.[44] Earlier, the continuance of the Gujarati-Portuguese war of 1546-49 drew anguished complaints from Portuguese officials and others. They pointed out that the war was costing the state and its inhabitants a considerable sum, and would continue to do so until it was ended and trade with Cambay was restored.[45] Portugal, in fact, could not afford to fight a long war with Gujarat.

Generally, a Gujarati request and complaint to the Portuguese was listened to, and usually acted on. As was noted in chapter three, in 1603 Diu was given a municipal council and the title of "city". The Portuguese residents, who were the sole voters for the new council, abused their new-found power, the Gujarati residents complained, and in 1613 the council and the title were revoked.[46] Changes in the rates of customs duties in Diu were made only after the resident Gujarati community had been consulted. If they agreed, the rates were raised.[47] On one occasion the Gujarati merchants of Goa got the Portuguese to delay the sailing of the Cambay trade fleet until they had got their goods on board, although it was neither easy nor convenient for the Portuguese to do this.[48] Other Hindus living in Portuguese areas were governed as regards matters of inheritance by local Hindu law, but the Portuguese judges agreed that this was a little harsh ("reguroso") for the vanias; they were allowed to be judged by Portuguese law, a privilege not extended to any other Hindu group.[49] There are extant several petitions from Diu's native population complaining of oppressive Portuguese officials and of overly zealous priests. In either case the usual result was a conciliatory letter to the Gujaratis from the Portuguese king and a reprimand for the offender, be he official or cleric.[50] On the most human level, this élite political position growing out of economic power was best summed up in a letter from a viceroy in 1644 to a judge in Bassein : "Nagagissa, a merchant residing in this city, is going to see about his sick wife in Diu, and as he is a person to whom I am obligated, I order you to give him all the help and favour that may be necessary".[51]

The reasons for the economic and political power of the Gujaratis become clear when we look more closely at the real Portuguese strength in sixteenth-century Asia. The motherland, then as now, was a small and poor country. The population was at most 1,500,000, and Braudel, after detailing its heavy imports of grain, says "It was a country rotten at the core, an enormous deadweight of which Philip was to find himself master in 1580. . .".[52] This meant that the Portuguese simply lacked the numbers, or the metropolitan financial backing, to achieve dominance in their colonies. Even where the proportion of Portuguese to Indians was highest, in the capital, Goa, the Europeans were outnumbered about five to one.[53] A Dutch observer noted how in Goa the Portuguese "dwell in the towne among all sorts of nations, as Indians, Heathens, Moores, Jewes, Armenians, Gusarates, Banianes, Bramenes, and of all Indian nations and people", and outside the town there were even fewer of them, especially as they were forbidden to become gancars.[54]

A closer look at Portuguese military strength is also revealing. This strength has been distorted by the contemporary chronicles, which are full of tales of a handful of Portuguese defeating with enormous slaughter hordes of "blacks". The Portuguese did have about fifty forts in Asia, and at least a hundred warships.[55] But they had no large areas, except turbulent Ceylon, and in fact a case can be, and was, made that they had more forts than they needed, or could afford. For any governor it was a crowning ambition to build a fort somewhere, or anywhere, and have his name carved for posterity above the entrance. Once built, a fort could never be abandoned, except in cases of most dire necessity. All too often precious money and lives were wasted on defending them, when on a cost-effectiveness basis more emphasis should have been given to the armadas.

At sea the Portuguese were not, after 1509, seriously challenged by any hostile fleet in western Indian waters until the arrival of the Dutch and English early in the seventeenth century. Yet this did not mean that they could enforce their claim to monopolize all trade in spices and control and tax all other Asian trade. First, the Portuguese never established a base off the mouth of the Red Sea. Secondly, the Portuguese administration was not as efficient as is, at least ideally, a modern one. Thirdly, and most important, it was logistically quite impossible for an imperial power with a home population of 1,500,000 people with no more than 10,000 able-bodied European and Eurasian soldiers and sailors in all Asia, and with a hundred warships, to control all Asian waters, and run the empire's economy without reference to native people. Like other, later, colonial powers, the Portuguese had no choice but to use, co-operate with, and conciliate economically important native groups.

NOTES

1. Elaine Sanceau, ed., *Cartas de Dom João de Castro* (Lisbon, 1955), pp. 263-4.
2. For these, and other, religious measures see "Provisoes a favor da Cristandade," HAG, especially ff. 2-6, 34-7, 47r-48, 73r-4r, 126r-32r; LM, XIV, 183-3r; "Provisoes dos Vice-Reis," HAG, III, 33-34; Assentos, II, 160-1; T.R. de Souza, "Xenddi Tax : A Phase in the History of Luso-Hindu Relations in Goa (1704-1841 AD)" in P.M. Joshi, ed., *Studies in the Foreign Relations of India* (Hyderabad, 1975), pp. 462-71.
3. Assentos, I, 431-7; for the meeting of 1632 when Hindus were present. Also see Assentos, III, 240-2; Fazenda, VIII, 207r-8; "Livro de Ordens Régias", HAG, I, 151-2r; "Livro de contrato e concepção da colecta", HAG, *passim*. It may be noted that usually decisions were made by the Conselho do Estado or the Conselho da Fazenda, both of which were small exclusive bodies. Most Christians also had little say in decision making. This was a reflection of the strict gradations of society in Portuguese India, for which see also chapter three.
4. APOCR, III, 540-2; "Provisoes a favor da Cristandade," ff. 28r-9r, 49r-50, 85r-6; "Senado de Goa—Cartas Régias", HAG, I, 95r; II, 66r-7, 88r-90, and unpaged letters of 1595-7; AHU, Codice 282, ff. 16r-7, 112. On the municipal councils in the Portuguese Empire see generally C.R. Boxer's *Portuguese Society in the Tropics : The Municipal Councils of Goa, Macao, Bahia and Luanda, 1510-1800* (Madison, 1965).
5. Two recent studies of the Saraswat Brahmins may be noted here; N.K. Wagle, "The History and Social Organization of the Gauda Saraswata Brahamanas of the West Coast of India," *Journal of Indian History* XLVIII (1970); Frank F. Conlon, *A Caste in a Changing World : The Chitrapur Saraswat Brahmans, 1700-1935* (Berkeley and New Delhi, 1977). Conlon's "Saraswat Brahmans" are a small sub-

division of the Gaud Saraswat Brahmin caste-cluster. In this chapter the people I call Saraswat Brahmins are part of the larger GSB group; my Goan informants identified them as such, but did not specify any further. They did, however, indicate that today most wealthy Saraswats in Goa are vaishnavite. The important Saraswat temple at Priol, near Ponda in Goa, known as the Manguesh temple, is shaivite, with Shiva, typically, being worshipped in the form of a lingam. In my seventeenth century lists of names the large number of Saraswats whose first names were Mangueshi or derivatives thereof leads me to suspect that shaivite Saraswats were in this period more important than is the case today.

6. J.H. van Linschoten. *The Voyage of John Huyghen van Linschoten to the East Indies*, 2 vols. (London, 1885), I, 229-30.
7. ANTTDR, XLVIII, 318r; LXI, 161-3r: "Senada Régias", HAG, III, n.p.; AHU, caixa 14, letter of Dec. 12, 1641. See also T.R. de Souza, "Glimpses of Hindu Dominance of Goan Economy in the 17th Century," *Indica* XII (1975), pp. 27-35. For a fluent discussion of a comparable topic, based on familiar published sources, see G.V. Scammell, "Indigenous Assistance in the Establishment of Portuguese Power in Asia in the Sixteenth Century," *Modern Asian Studies* XIV (1980), pp. 1-11, and for an overview of the role of "collaborators", Ronald Robinson, "Non-European Foundations of European Imperialism : Sketch for a Theory of Collaboration," in Roger Owen and Bob Sutcliffe, eds., *Studies in the Theory of Imperialism* (London, 1972), pp. 117-42. Dr. de Souza makes several useful corrections to my earlier "Indigenous Dominance in a Colonial Economy," which is subsumed in the present chapter. I have as usual benefited from his comments, but in this article, and in his later *Medieval Goa* (Delhi, 1979), p. 146, his view about who was meant to be forced to marry Portuguese is questionable. The relevant quotation refers to "nenhũ Bramane, nẽ charado, ou de outra qualqer casta q seja sendo riqos ou tendo fazas."
8. This part of the chapter could not have been written without the generous help of Mr. G. Ghantkar late of the Historical Archives of Goa, who gave me information on the castes of the people holding rendas, and on the modern spelling of their names. Information for rendas from 1600 to 1670 comes from the following volumes in HAG : "Arrematação das rendas do Estado," I, *passim*; "Termo das Fianças", I-III, *passim*; "Provisoes dos Vice-Reis", I, 164-76r; 193-4, II-III *passim*; "Alvarás e Provisoes de Sua Magestade", I, 318r; III, 17; Fazenda, I-X, *passim*; LM, XIV, 22; XVIII, 81; "Correspondencia de Baçaim", I, 45; "Cartas e Ordens", III, 24r, 27, 57, 81r, and from Pedro Barreto de Rezende, "Descripçoens das Cidades e Fortalezas da India Oriental", II, 150-1 of the copy in the Biblioteca da Académia das Ciências, Lisbon; ANTTC, 3-27-44; ANTTDR, LVI, 386-88r; AHU, India, caixa 7, relação of March 20, 1623; caixa 8, treslado of Feb. 19, 1625; AHU, codice 219, ff.

4-54. For the rendas of 1528, 1541-43, and 1702-5 see ANTTCC, 3-10-75; "Cartas dos Vice Reis" ANTT, no. 134; "Arremetação das rendas do Estado", HAG, III, *passim*. For the smaller contracts of 1654-64 see "Livro de Termos das Arremetaçoes das Rendas", HAG, I, *passim*, and "Senado de Goa. Termos das Obras", HAG, I, *passim*.

9. Luis de Figueiredo Falcão, *Livro em que se contêm toda a Fazenda e Real Patrimonio dos Reinos de Portugal, India e Ilhas adjacentes* (Lisbon, 1859), pp. 75-78; AHU, India, caixa 6, account of Aug. 19, 1618; Barreto de Rezende, II, 150-3r; "Livro de registo dos alvarás, cartas etc. de diferéntes feitorias," HAG, I, 16-30.
10. Fazenda, IV, 170; V, 50r; IX, 38-40; LM, XIIIA, 161-79; Ali Muhammad Khan, *Mirat-i Ahmadi*, trans. M.F. Lokhandwalla (Baroda, 1965), p. 193.
11. AHU, India, caixa 2, no. 252.
12. For Diu see Barreto de Rezende, I, 58r; APOCR, VI, 1250-1; LM, XXVIIIB, 419, 422-6; ANTTDR, LI, 131-1r, 136-6r AHU, India, caixa 14, petition of Sept. 10, 1642, *et seq*. For Hurmuz, "Livro Morato", HAG, ff. 331-1r; for Goa, Fazenda, VI, 13lr-2; "Chancelarias Reaes", ANTT, Affonso VI, livro 39, ff. 372-2r.
13. See "Livro das termos das fianças", HAG, I, *passim*.
14. LM, XXVIB, 407; Barreto de Rezende, II, 121. T.R. de Souza ("Glimpses", pp. 29-30) finds that the Hindu population declined during the seventeenth century.
15. *Cartas de Affonso de Albuquerque*, 7 vols. (Lisbon, 1884-1935), I, 306.
16. LM. XIV, 249-323r.
17. AHU, caixa 18, letter of Jan. 25, 1646.
18. Fazenda, VIII, 182; IX, 247r-8r; AHU, caixa 4, no. 109; J.H. van Linschoten, I, 228.
19. LM, XXVIB, 408.
20. B. Mus., Add. 28433, f. 116r.
21. Figueiredo Falcão pp. 75-8.
22. Couto, V, ix, 1.
23. António Bocarro, *Decada 13 da História da India* (Lisbon, 1876), p. 337.
24. Barreto de Rezende, II, 125-8r.
25. Ibid., II, 128r; François Pyrard de Laval, *The Voyage of François Pyrard de Laval*, 2 vols. (London, 1887-90), II, 246; John Jourdain, *The Journal of John Jourdain* (Cambridge, 1905), p. 173.
26. LM, XIIIB, 392-415r.
27. Caesar Frederick in Richard Hakluyt, *The Principal Navigations, Voyages, Traffiques and Discoveries of the English Nation*, 12 vols. (Glasgow, 1903-5), V, 377; J. Wicki, ed., "Duas Relaçoes sobre a situação da India Portuguesa nos Anos 1568 e 1569" *Studia* VIII (1961), p. 177; Artur Viégas, ed., *Relação Anual das Coisas que*

Fizeram os Padres da Compánhia de Jesus nas suas Missões, by Padre Fernão Guerreiro, 3 vols. (Lisbon, 1930-42), II, 389; Barreto de Rezende, I, 63r-64.

28. *Studia*, XIII-XIV, p. 86; AHU, caixa 18, letter of Jan. 25, 1646.
29. M.N. Pearson, *Merchants and Rulers in Gujarat* (Berkeley, 1976), p. 23, and footnote 65.
30. Pyrard de Laval, II, 246; Jourdain, p. 173; Frederick in Richard Hakluyt, V, 375-6.
31. AHU, caixa 17, petition of Aug. 20, 1644.
32. LM, XXVIIIB, 352; "Livro Morato", HAG, ff. 262-2r.
33. Fazenda, II, 25r-6; cf. APOCR, I, ii, 213.
34. Assentos, I, 431-7.
35. "Livro Vermelho", HAG, I, 55-6.
36. "Senado Régias", HAG, II, n.p.; "Regimentos e Instruçoes", HAG, II, 10r-11r.
37. Assentos, III, 395-8; Diary of Viceroy Sarzedas in "Noticias dos Estados da India," Biblioteca da Académia das Ciências, Lisbon, Mss. Azuis, A58, f. 48.
38. See works cited in footnote 37 of chapter two.
39. AHU, codice 281, ff. 23-23r; "Alvarás e Provisoes de Sua Magestade", HAG, I, 218r; António Bocarro, *Livro das plantas*, in A.B. de Bragança Pereira ed., *Arquivo Português Oriental*, 11 vols. (Bastorá, 1936-40), tomo IV, vol. II, part 1, p. 310; Assentos, II, 12-13, 54-56, 167-9; Fazenda, VIII, 99r, 146r, 186r, 194; LMBP, V, 145.
40. AHU, caixa 5, no. 6; Fazenda, I, 62r-3r; ANTTDR, XXIII, 307; XXIV, 31-1r; LMBP, IV, 18.
41. "Provisoes dos Vice-Reis", HAG, I, 96.
42. For example, Correa, III, 419; Couto, VII, iii, 3.
43. See *New Cambridge Modern History* (Cambridge, 1957-68), II, 610-11, III, 536-9.
44. Bocarro, *Decada 13*, p. 336; AHU, caixa 3, no. 116; LM, XII, 216-6r.
45. "Colecção de São Lourenço", ANTT, III, 185, 189, 480-1r; ANTTCC, 1-81-104.
46. AHU, codice 282, f. 126; LMBP, IV, 22, 538-40, 604-5; ANTTDR, XXIV, 23r; B. Mus., Add. 20861, f. 89r.
47. LMBP, III, 157; B. Mus., Add. 20861, f. 59r; Biblioteca da Ajuda, Lisbon, 51-VII-30, f. 97; AHU, caixa 10, report of Sept, 30, 1634.
48. Fazenda, VI, 131r-32.
49. "Alvarás e Provisoes de Sua Magestade", HAG, II, 16-20r; APOCR, VI, 1267-73; AHU, India, caixa 16, parecer of March 14, 1644, and caixa 19, parecer of Feb. 13, 1647.
50. For example, "Colecção de São Lourenço", ANTT, V, 133; LMBP, II, 30, AHU, India, caixa 14, petitions and letters of Sept. to Dec. 1642; "Livro Azul", HAG, II, 41r-43r; ANTTDR, LIV, 220; LV; 196: LM, XXVIIIB, 419; APOCR, VI, 840.
51. "Correspondencia de Baçaim", HAG, f. 88.

52. C.R. Boxer, *The Portuguese Seaborne Empire* (London 1969), pp. 52-3; Fernand Braudel, *The Mediterranean and the Mediterranean World in the Age of Philip II*, 2 vols. (London, 1972-73), I, 585-91.
53. Barreto de Rezende, II, 121; "Livro Verde", HAG, I, 225.
54. J.H. van Linschoten, I, 181-2.
55. For seventeenth-century problems with men and supplies see A.R. Disney, *Twilight of the Pepper Empire*, pp, 63-70.

INDIAN SEAFARERS IN THE SIXTEENTH CENTURY 6

> Then came the news that Mrs. Tulsi had decided to send Owad abroad to study to become a doctor. . . And at Arwacas some of Mrs. Tulsi's retainers defected. Forgetting that they were in Trinidad, that they had crossed the black water from India and had thereby lost all caste, they said they could have nothing more to do with a woman who was proposing to send her son across the black water. 'Water off a duck's back', Mr. Biswas said to Shama. 'The number of times that mother of yours has made herself outcast!' There was talk about the suitability and adequacy of the food Owad would get in England. 'Every morning in England, you know,' Mr. Biswas said, 'the scavengers go around picking up the corpses. And you know why? The food there is not cooked by orthodox Roman Catholic Hindus'.
>
> (V.S. Naipaul, *A House for Mr. Biswas*)

MY AIM in this chapter is to investigate a little more closely than has been done previously exactly who went to sea in the sixteenth century. Some of the flexibilities (and confusions) about sea travel and food evident in the above passage from V.S. Naipaul appear in the sixteenth century data also. Perhaps for this reason, this aspect of India's maritime history has been largely ignored.[1] Thus Qaisar's excellent paper[2] contains useful information on ship building, and on the internal management of ships, and generally uses very scattered Persian sources most competently, but the author has nothing to say on the communal composition of the seafarers. Dr. Varadarajan[3] has written a very important paper, of great interest for the data in it, and also methodologically, because it is based on oral tradition. It is true that there are problems in using late twentieth-century oral tradition for earlier periods, but a combination of her technique with archival research would be most fruitful. She mentions various caste and Muslim groups who traditionally have been seafarers and stresses their interaction. But she makes no attempt to explain how her Hindu sailors, not all of them low caste, can justify sea travel. As maritime history develops we can hope for many more such papers.

This chapter is concerned with the communal composition of Indian seafarers in the sixteenth century. I will first discuss briefly the consensus of scholars on the situation at 1500, and try to add a little detail so that we can understand more fully the situation at this time. I will then investigate changes in the communal composition of western India's overseas traders and travellers in the sixteenth century. It is clear that Hindus of many castes travelled extensively by water at this time. I will try to explain this phenomenon at the conclusion of the chapter.

The bulk of the chapter studies people who travelled on the sea. Nevertheless, I think it is a comment on the whole notion of "maritime" history that problems appear when we consider the theoretical position, for the *dharmasastras* discouraged sea travel. Here then land matters, in this case brahmanic injunctions, impinge quite directly on what happens on the sea. Further, the solution to this problem of a gap between theory and practice seems also to come from a study of "internal" medieval Hindu attitudes and values. Not that I claim to have solved the problem, but I am convinced that the answer lies somewhere in the areas sketched at the end of the chapter.

It turns out to be extraordinarily difficult to find conclusive data on who went to sea. The authorities are sometimes in error: thus Auguste Toussaint's standard *History* claims, incorrectly, that the famous Kunjali Marakkars of Malabar were Hindus.[4] The Portuguese sources on which we must primarily rely for information on the sixteenth century are often confused. As we will see, a "baneane" (or other variants) is a vania or banya, but apparently a "brahmene" is not always, as we would expect, a brahmin. The word, it seems, was sometimes used to mean Hindus in general, like the more usual Portuguese term "gentio." When we chance on the actual name of a sea trader in the Portuguese records it is often mangled to the extent that I at least have trouble in correct reconstruction; the appendix to this chapter will illustrate this.

It is crucial to keep in mind that a merchant who lives by overseas trade and resides in a port does not necessarily ever go to sea himself. A sea trader does not have to be a sea traveller, a sailor, or a navigator.

What sorts of people would be likely to be on a ship in the Indian Ocean in the fifteenth and sixteenth centuries? One person usually not on board was the owner or owners of the ship. It seems clear that they were usually shore based. However, sometimes, the owner was also the captain, or *nakhoda*, of the ship. Two points need to be made about this person. First, the nakhoda was not the actual navigator, who was called the *sarang*. Second, while the nakhoda was usually a private individual, on the Red Sea route on the pilgrim ships he was a Mughal official and mansabdar. Like everyone else on board, the nakhoda was also a trader. Thus Barros noted how the Portuguese in 1516 captured two fine Gujarati Muslim ships. The "captain" of the two ships was also the owner of the greater part of the goods on board.[5] The actual sailors, with their headmen, were not salaried employees of either the owner or the nakhoda; but also lived from trade. However, as Abul Fazl gives us salaries for all the crew on a ship, presumably sometimes payment was partly in cash and partly in concessions concerning free carriage of goods. The passengers were mostly small traders with one or two bundles of goods. Some acted as agents for shore based big merchants, but most of these ubiquitous petty traders were autonomous and independent. Professor Arasaratnam says that the petty trader, the pedlar of Van Leur and Steensgaard, was less conspicuous by the seventeenth century, by which time most trade was done by nakhodas on behalf of their shore based principals. Finally, ships travelling to the Red Sea would be likely to have some or many pilgrims on board. Here again categorization is difficult, for many pilgrims traded enthusiastically, and returned, as one sour English observer said, "freighted with Religion and Pelf".[6]

Our authorities are agreed that shipping in the Indian Ocean in the fourteenth and fifteenth centuries was dominated by Muslims. Moreland at first depicted Arab dominance, but later modified this to "Moslem." He saw Muslims, local and foreign, based in Bengal and Gujarat, as important participants in sea trade. His conclusions are worth quoting rather fully : da Gama

> found the Indian Seas from Madagascar to the Straits of Malacca practically in possession of the Moslem merchants, who owned and managed most of the ships and also took an important share in the trade in land. Traders of other classes could hire space on these ships for cargo, and could travel with their goods, but they had practically no influence on the shippers other than what was derived from their demand for cargo-space, except in the cases where they owned the ships for themselves. Such cases were comparatively rare : from the accounts given by Barbosa and Varthema I gather that the Moslems controlled practically all the ships on the Malabar coast, the great majority of those plying from the Gulf of Cambay, and a large share of those on the Coromandel coast and in the waters of Bengal. Whatever numbers might be owned by Bengali, Coromandel or Gujarati merchants, the predominance of the Moslem interest is beyond dispute. . .[7]

This broad outline has been generally accepted as a true statement of the position at 1500.[8]

It is true that a few scholars, including Moreland himself, also mention in passing trade by Hindus, especially Gujarati vanias and Coromandel Tamils.[9] Nevertheless, Muslim dominance is generally accepted. It is explained by describing how the Muslims slowly converted the coastal people, including merchants, with whom they came into contact, a process which seems to be well attested not only in India but also in Southeast Asia and East Africa. As a sort of mirror image of this conversion, those who did not convert became less willing to travel by sea. As Simkin put it, Arabs and Muslims displaced Hindus "partly because of a new, Brahmanical ruling that Hindus sinned by travelling on the sea".[10] This religiously based explanation for what is seen as a decline of Hindu trade is given most eloquently by A.L. Basham :

> Though maritime trade did not cease it became more and more the affair of foreign merchants. Indians never wholly abandoned the sea, but by the time of the Muslim invasion travel to foreign lands was believed to bring grave impurity

> upon members of the upper classes, and this, according to some authorities, could never be expunged. This religious objection to sea travel was a measure of the growing fear of and distaste for the sea, which in some degree existed at all times. Though Indian mariners were resourceful and by no means lacking in courage, sailing the sea was always depicted as hateful and desperately perilous. By the Middle Ages the Arabs and Chinese had outstripped the Indians in the art of ship construction and it became more profitable for the merchant to sell his wares to foreign middlemen than to take them abroad himself. The Muslim invasions encouraged xenophobia, and the people who had planted their colonies from Socotra to Borneo became, with religious sanction, a nation of landlubbers.[11]

How much truth is there in this depiction of Muslim dominance and Hindu decline or even defeat? We need first to add some details to the picture at 1500, and then talk of sixteenth-century changes. It is true that the accepted picture of Muslim dominance is broadly true, but there are stray indications of Hindus and Jains at sea also. Thus in the early fifteenth century on a becalmed ship the sailors performed a rite to the god "Muthia" to get wind.[12] This is the well known Tamil god Muthiah, and the sailors concerned were clearly Hindu. Similarly, Barbosa mentions several groups of lower caste Hindu sailors and fishers.[13] More generally, one's suspicions are aroused by the rather frequent mentions of Hindus and Muslims trading all over Southeast Asia in 1500, or of the Portuguese taking ships and killing all the Muslims on board. The rest of those present were obviously non-Muslim, and presumably Hindu.[14]

A brief survey from East Africa to Southeast Asia will help to flesh out a picture of Hindus trading and settling rather extensively around 1500. At Melinde we hear of locals trading with "the Muslims and Hindus of the kingdom of Cambay who come to this port with many ships. . .".[15] Da Gama found at Melinde "many Hindu Gujaratis" who lived and traded there.[16] Serjeant sees the Arabs as the aristocracy of East Africa in 1500, but finds also many Indian residents and traders, most of whom were not Muslim.[17]

Leaving aside the area from the Red Sea to the Persian Gulf, which, except for Hurmuz, at 1500 appears to be virtually totally Muslim, we come to Gujarat. The numerous Portuguese descriptions make it clear that by 1500 foreign Muslim merchants were less dominant than previously, and that Hindus and Muslims both traded extensively. The general picture seems to show Hindus and Jains dominant on land, owning most of the capital, and presumably many ships. We hear of thriving, prosperous non-Muslim merchants with an accepted and honoured position in Gujarati society.[18] The question is how many of these Hindus and Jains went to sea? Obviously those settled in East Africa and all up and down the west coast of India had got there by sea, and may well have returned home to die. Nevertheless, Malik Gopi was probably more typical. He was a major ship owner and trader in the early sixteenth century, but as a Nagar Brahmin did not travel by sea himself. Typically, he used Muslim seamen for his ships.[19] Albuquerque noted more generally "the vanias of Cambay, who are the main gentile (sc non-Muslim) merchants of these parts, whose ships' crews are all Muslim".[20] The actual sea traders were more often Muslim than Hindu. The well-known situation in Malacca bears this out. Most of the important Gujaratis there were Muslim; Hindus were also influential, but they came from Coromandel.[21]

Nevertheless, there is also powerful evidence from the well informed Tomé Pires which tends to show at least some non-Muslim Gujaratis travelling by sea. The passage in question is not completely free of obscurities. It says "the heathen of Cambay—and in olden times the Gujaratees—held that they must never kill anyone, nor must they have armed men in their company. If they were captured and [their captors] wanted to kill them all, they did not resist. This is the Gujarat law among the heathen. Now they have many men-at-arms to defend their ships".[22]

This passage is particularly valuable because it does differentiate communally. In other places Pires, and others, refer simply to Gujaratis, or merchants of Cambay. They also frequently talk of merchants trading to various places both by land and sea; unlike the passage quoted above, there is no clear statement as to who sailed in the ships. A possible explanation for this usual failure to differentiate is that Hindus and Muslims did mingle and trade together, and to an undiscerning observer there may not have appeared to be much difference, especially as many of the Muslims were local converts. The degree of mixing and cooperation is perhaps best seen in Zain-ud-din's passing reference to a Mlabaris seizing ships belonging to the "pagans" of Gujarat and the Konkan. He adds that these local ships were usually jointly owned by several people, and sometimes one of the joint owners was a Muslim.[23]

Moving south from Gujarat in our survey, we come to the Konkan area. In ports like Chaul and pre-Portuguese Goa it is clear that at least long distance sea trade was in Muslim hands. Goa in fact was notorious for its Turkish inhabitants, a fact which contributed to Albuquerque's desire to conquer the city. It is, however, worth remembering that one person who was influential in encouraging Albuquerque to this attack was the Hindu Maratha corsair, Timoja. There was also a rich sea-trading community of chatims based on Barcellor when the Portuguese arrived.[24]

The situation in Malabar at 1500 can also be sketched rather quickly. The sources are united in their depiction of foreign Muslim dominance of the overseas trade of the area, and especially in Calicut. Arab merchants from the Red Sea and Egypt were most conspicuous.[25] It appears, however, that non-Muslims, and especially Gujaratis, handled much coastal trade to the north. They were certainly prominent and numerous in all the ports of Malabar, especially Calicut.[26] As in the case of Gujaratis in East Africa, even if these people did not trade on the sea they certainly had travelled by sea to get to Malabar. In fact some coastal trade around to Coromandel was also handled by Hindus, for Barbosa noted a mountain in the extreme south "whither all the ships of both Moors and Heathen steer. . .".[27] But generally the sources agree, to quote Pires, that "All the merchants in Malabar who trade on the sea are Moors and they have the whole of the trade." Similarly, Varthema said "It must be known that the the pagans do not navigate much, but it is the Moors who carry the merchandise. . .".[28] Pires, however, contradicts himself and points to one group of Malabari Hindus who went to sea, and not low caste ones either. He notes that certain Nayars travelled by sea, but only "in case of dire necessity", but then makes it appear common practice by claiming that sea merchants in general in Malabar "have paid Nayars who accompany them; and some of these Nayars are their secretaries and are better accountants than the Moors".[29]

It is evident that at 1500 the Hindu Indian groups most extensively engaged in sea trade were those from Coromandel, especially Chettyars. Barbosa noted the "chetijs," and their ample girths, in Malacca.[30] Hindu trade from Malacca seems to have been focussed on Coromandel and Malabar. Longer distance trade was controlled by Muslims from the Middle East, Gujarat, Bengal and also Coromandel.[31]

The broad picture of Indian Ocean commerce at 1500 remains one of Muslim superiority, though not total dominance. Hindus from Gujarat and Coromandel were the two groups most important among non-Muslim sea traders. Before considering changes in the sixteenth century, we can look at the available information on who crewed these ships. All our information shows that Muslims were dominant in this activity. The Portuguese recognized the skill of Muslim pilots. In 1541 they hired two Muslim pilots "who knew sailing conditions in the Red Sea".[32] Earlier Pires paid his tribute: "The Gujaratees were better seamen and did more navigating than the other people of these parts, and so they have larger ships and more men to man them. They have great pilots and do a great deal of navigation".[33] These Gujarati sailors were often *navayats* ("naiteas") from the port of Rander.[34] There are several notices in Portuguese records of the navayat sailors, their headmen, and their skill and value to the Portuguese.[35] It is true that the Portuguese also used Arab and Kanarese sailors, and indeed by number the last probably dominated their crews.[36] There is no indication in the sources of the religion of these Kanarese : presumably they were Muslim, and at least some were Kanarese-based navayats. On Portuguese-owned ships the captain was Portuguese, but on Indian-owned ships he was probably a navayat. In the appendix I list some valuable information from the Goa records of some passes granted in the early seventeenth century. The numerical dominance of Muslim nakhodas, and a fair sprinkling of navayats, are two points to emerge from this list. These passes, were, however, nearly all for ships belonging to the political élite of western India. More typical of merchant shipping, but still showing the importance of navayats, is a pass issued in 1627 to the ship "Salamoty" of 400 candis to go to Dofar. Her owner was Tacu Chatim, a married man living in Goa, but her captain and nakhoda was "Xilema," a Muslim of the "naitea caste," and her crew of forty were of the same "caste".[37]

Moving further south, it is clear that Muslims sailed most ships operating from Malabar. Those controlled by the Kunjali Marakkas appear to have been crewed by the Kunjalis' fellow Moplahs. However, there is more than one mention of fighting Nayars on board ships, acting not as sailors but as marines, or, as noted, as secretaries.[38] Other Hindu groups of lower castes were widely used in southwest Indian waters as common sailors and oarsmen. These groups were often fishermen and sailors by hereditary occupation, and were close to the bottom of the Malabar Hindu hierarchy. These seafaring castes included Mukkuvars, Tivars, Mogers and Pulayars.[39]

We have now to reverse our direction and travel up the west Indian coast, surveying as we go the fate of various sea trading people during the sixteenth century. The situation in Malabar is relatively well known.[40] The Muslim trading community had been divided into foreign and local Muslims, the former hailing from the Red Sea area, Egypt, and Turkey, the latter mostly Moplahs. The Portuguese had been able to play on these differences, and this, coupled with a sometimes ambiguous attitude from the Zamorins, meant that the foreign merchants decided to leave. Many of them were killed by the Portuguese on their way to safer parts. The local Muslims did not of course have this option of flight : they had to stay and made the best of it. The struggle of the Kunjali Marakkars against the Portuguese for the whole of the sixteenth century has been well described by Professor Nambiar and also the contemporary Zain-ud-din.[41] Other Malabar Muslims simply tried to keep on trading. Richer, or more ambitious, merchants in the 1520s and 1530s at least still tried to send ships to the

Red Sea. Others traded in pepper north to Gujarat, and so to the vast north Indian market for spices. The Persian Gulf area also provided a market for Malabar pepper. Some of these traders were also part-time pirates, attacking any other merchant ship they came upon; it was to meet this challenge that the Portuguese started their system of convoys, or *cafilas*.[42]

As in 1500, some of the sailors, soldiers and rowers on these trading, corsair and pirate ships were Hindu, but the captains and most of the soldiers were apparently Muslim. Indeed, as we noted in chapter two, as far as the Portuguese were concerned the word "Malabare" was synonymous with "pirate" and with "Muslim." However, the links we noted earlier between Hindus and Muslims in Gujarat were also in evidence in Malabar. The Zamorins, and the Nayar warriors, cooperated at times with the Moplah corsairs and traders. As we noted in chapter two, the organization for harvesting and transporting pepper in Malabar involved the Nestorian Christians, Hindus and Muslims at different stages. But the sea-going side of this attempt to defy Portuguese control was handled almost completely by Muslims.

The Malabaris operated all up and down the west coast, including the Kanara and Konkan areas around Goa. In this region also the Portuguese presence had an effect. As in Malabar, some local traders accepted the Portuguese system, participating in the convoys of rice, several of which came north to Goa each year. The crews no doubt came from the same presumably Muslim Kanarese communities who manned Portuguese ships, though the owners and financiers were likely to be local Kanara Hindus. The chatims of Barcellor lost most of their sea trade, thanks to Portuguese opposition, but they continued to defy the Portuguese by creating disturbances when Portuguese ships came in to Barcellor. They also took pleasure in supplying rice to Malabari traders rather than to the Portuguese.[43] The tradition of Hindu piracy, to which the Maratha brothers Timoja and Raoji, based on Honawar, belonged, was continued in opposition to the Portuguese by the naik of Sanguicer. This naik had been loosely subordinate to Bijapur, but had revolted. He built light armed ships which carried twenty fighters and terrorised merchant shipping protected by the Portuguese all up the coast north to Goa. Indeed, this type of ship was named after the river where the naik had his base, becoming known as "sanguiceis." He was defeated by a joint Portuguese-Bijapuri force in 1585,[44] and I have found no further mention of Hindu pirates in the sixteenth century.

It is unclear how many Hindus from the central west coast of India travelled by sea, for the sources tell us little. However, there is copious evidence of Gujarati vanias, mostly Hindu but possibly also Jain, travelling and trading all over the west coast from Malabar north. Thus a letter of 1598 refers to vanias, natives of Gujarat, trading in their own ships to Chaul, Goa and Cochin.[45] Nor were they averse to travelling in Portuguese ships from Goa to Cochin.[46] Similarly, Portuguese records mention advice received from one vania or another who had recently arrived from some other port : thus the Portuguese in Bassein got news from vanias just in from Surat, Diu, and Cambay, for example.[47] In 1564 Portuguese pirates seized a merchant ship going from Diu to Cambay, cut off the hands of all the vanias on board, and threw the unfortunates overboard.[48] Nor were vanias only engaged in coastal trade. In 1626 a Portuguese ship left Goa for Mossambique. Among those on board were four Portuguese merchants, four mesticos, and four local Christians. There were also three vania goldsmiths ("baneanes oriues"), by name Roulo Chatim, Purcia Chatim, and Bua Chatim, all of whom lived in Goa.[49]

In the later sixteenth century the main coastal trade for vanias was that from the Gulf of Cambay to Goa, carried in convoys under the protection of Portuguese warships. The volume of trade on the route rose greatly, for in the fifteenth century it had not been important for Gujarati traders. Three variants may be noted. First, in many cases merchants themselves travelled on these small coastal craft, staying in Goa only as long as they needed to complete their business.[50] Second, others of these itinerants acted as agents or factors for bigger merchants based in the ports of the Gulf of Cambay. Third, Portuguese records differentiate clearly between vanias visiting Goa, or attending at Goa, and those who were residents and householders. The former in fact were subject to Gujarati law, the latter to Portuguese.[51] It may be noted again that even the vania residents in Goa, and all the other ports of western India and indeed seaborne Asia, must have travelled by sea at some time to get to their new home. The picture in Goa is thus of streets full of vanias, some resident and some visiting. Portuguese records made clear their economic importance to the Estado da India, and consequent Portuguese concessions, such as the legal one just mentioned, and their right to have their own "captain". The captain was presumably a resident. We have the names of two of them from the late seventeenth century (Raichandrassa Vagassi, and Odo Beddu Chatim) but it clearly was not then a new position.[52]

Two other points can be made about these seafaring vanias. First, trade was not the only reason for sea travel. "Nagagissa" once went from Goa to Diu to help his sick wife. We also have a case of a female vania, "Giui Baby," coming to Goa on legal business.[53] Second, it must be noted again that the stress on communal divisions on which this chapter focusses should not be allowed to disguise the degree of Hindu-Muslim cooperation which existed. The crews of the Gujarati ships were probably navayat Muslims, although the owners and traders were usually Hindu, if only because the Portuguese preferred them to Muslims. Similarly, in 1620 the Muslim governor of Cambay had a vania, "Mulam Peros," as his agent in Goa.[54]

To conclude this discussion of Gujaratis in western India, we need to consider whether other Gujarati Hindus apart from vanias travelled by sea. In 1583 we hear of a resident of Diu called "Serepte," a rich brahmin ("hum Bramene rico").[55] In the early seventeenth century Pyrard noted that in Goa "the goldsmiths, lapidaries, and other workmen occupied with the finer crafts are all Banians and Bramanis of Cambaye".[56] Finally in Goa in 1634 the "brahmin natives of Gujarat living in Goa and Diu" ("bramanes naturaes de guzarate moradores na Cidade de Goa e em dio") petitioned the Portuguese authorities. The case dragged on, and in 1638 the vanias ("baneanes") asked for a decision.[57] One could assume that the "bramanes" of 1634 are only a copyist's mistake for the real petitioners, the "baneanes" of 1638, except that "bramanes" appears in two separate copies, in different archives. As for Pyrard, he had earlier noted an interpreter in Cochin who was "a Banian and Brameny, by race and religion. . .".[58] His editor says that for

Pyrard "brahmeny"simply meant Hindu. While the Goa goldsmiths could have been brahmins, it is more likely that here also Pyrard means they were vanias by "race" and Hindu by religion.

Pyrard's evidence is thus confused. However, one would expect the numerous Gujarati vania overseas communities to employ their own brahmins. The Portuguese sources are almost silent on this crucial matter, but this conjecture is strongly supported by the finding of Prof. Ashin Das Gupta that in the eighteenth century in Mocha vanias "brought their Brahmins with them. On days which required public celebration of religious rituals they, with their priests, would go outside the town walls and hold their ceremonies under the sky. Temples were not built, but several houses in town were used for worship. . .."[59] Two hundred years earlier, Jean Aubin tells us, brahmins were to be found in Hurmuz.[60] It is also significant that Tomé Pires noted of Malabar early in the sixteenth century that the "Brahmans go to sea even less" than the Nayars;[61] by implication some brahmins *did*. On the other hand the Portuguese in 1500 offended grievously by detaining high-caste people on board ship, for this meant they could take no food.[62] It is clear enough that in the medieval period some brahmins could travel by sea ; it would be fascinating to know exactly what sorts of brahmins these were. They may well have been Saraswats, for three centuries later the *Bombay Gazetteer* noted of this group that the majority "allow widow-marriage, and freely travel across the seas to collect payments from Kachh Vania traders in Mozambique and Arabia".[63]

We must now look west from Gujarat and consider vania activities in the Red Sea and Persian Gulf areas. As usual our evidence is more scattered than we would like, but certain changes seem to have occurred. There seems to be good evidence that in the sixteenth century long distance overseas trade from Gujarat was dominated by Muslims resident in Rander and Diu. Cambay, Surat, Gogha, Broach and the ports west of Diu no doubt all had important traders too, and of course the Gujarati goods exported from these two ports came from all over the kingdom, and even further a field. But we may suspect that at least some Hindus also took part in this trade, for S.D. Goitein finds evidence from the Cairo geniza documents of many Hindu merchants in Cairo in the twelfth century.[64]

Rander, on the right bank of the Tapi river, was of course the home of the navayat Muslims we have mentioned before. A reliable description of the port at about 1520 said it was a very fine city with extensive trade to Bengal and all over Southeast Asia. Rander Muslims sailed large beautiful ships and themselves became rich and honoured. Malacca and China goods were cheaper and more available in Rander than anywhere.[65] These Muslim ships were the ones which the Portuguese in their early years attacked so strenuously. But, on the other hand, these were the Muslims who often fought back valiantly, or even attacked the Portuguese without direct provocation.[66] Rander ships often traded for years before returning to Gujarat. Nor was their trade only to the east, for Barros describes the port as being the centre of Gujarat's trade to the Red Sea also.[67]

Diu, the other major port, under the astute control of Malik Ayaz, also traded extensively to the Red Sea and elsewhere. Many Muslim merchants from the Red Sea visited there, as did numerous Turks. It seems to have been, like Rander, a port dominated by Muslims, but, unlike Rander, most of the Muslims in Diu were foreign.[68]

Both ports suffered grievously at the hands of the Portuguese in the 1530s. Diu was constantly harassed, and finally the Portuguese were able to build a fort there in 1535. Five years earlier, along with other Gujarati ports, Rander had been most thoroughly sacked.[69] The result was a decline in long distance Muslim shipping from Gujarat. Some, especially the foreigners, presumably moved elsewhere to an area less closely controlled by sea than Gujarat. Others traded under the Portuguese system, especially to the Red Sea.[70] The navayats are mentioned as revolting against a local noble in Rander in 1560,[71] but they too seem to have in effect submitted, becoming, as we have noted, relatively (in Portuguese eyes) law abiding sailors and pilots. Some Gujarati Muslims, however, continued to probe the Portuguese system for weak points. In particular, in the 1540s there was a flourishing "illegal" trade in spices from the Bay of Bengal to Gujarat and especially the Red Sea. This trade was dominated by Gujarati Muslims, and Turks, the former according to one witness being from Rander and Surat.[72]

The people who accommodated best to the Portuguese system, and prospered in its entrails, were the Gujarati vanias who lived in, or moved to, Diu. This is not to deny that there continued to be Muslims resident in, and trading from, Diu throughout the sixteenth century. Nor should it be forgotten that in all likelihood the crews of the ships owned by vanias were Muslims. Nevertheless, the dominant position of vania capital in Diu, and their leading role in financing its huge overseas trade in the late sixteenth century, is well attested. Some, it is true, were only agents for other vania merchants based on Cambay, Surat and other ports, but many were independently wealthy. These Diu vanias had their own "captain," a powerful figure in Diu politics and one whom the Portuguese had to conciliate.[73]

More specifically, a petition of 1642, possibly self-serving, from the vanias of Diu complained about a priest who was forbidding them to build temples. As the petition said, they or their ancestors had come to live in Diu "in the shadow of your Majesty and his Viceroys, and with our trade we have enriched this custom house, for the merchants of this area being all gentiles, and Muslims and Christians very few, your Majesty should know well that unless we can do our rites and ceremonies and customs in our temples we will not be able to stay in Diu."[74] Further concrete evidence comes from a Jesuit account of about 1600. It describes how formerly the trade from Gujarat to the Red Sea had been done via Diu by Red Sea merchants. In the last few years vania merchants resident in Diu, or agents for others, had almost taken over this trade. The Red Sea merchants acquiesced because this meant less trouble and danger for them, and so greater profits. But the change was bad for the Portuguese, because the vanias were expert at cheating on customs. Further, the Portuguese in Diu were now coerced by the vanias to block anyone else from trading to the Red Sea. Even the clergy were dominated by them.[75] This account is most valuable, but it must be stressed that while the vanias now did much better, they had participated in this trade already for several centuries.[76]

The prosperity of these vanias and Diu was interlocked, and when one declined so did the other. Thanks to Dutch and British attacks or at least competition, Portuguese fortunes collapsed in the seventeenth century. Even by the 1630s Bocarro noted how the trade of Surat, protected by the Dutch and British, had usurped Diu's role. The situation for the Portuguese deteriorated progressively from then on. Where once there had been 200,000 rich and powerful vanias in Diu, by the 1660s there were only 7,000 or 8,000 poor ones. The customs house once yielded over Rs. 1,50,000 a year, but in 1667 only Rs. 54,000.[77] The vanias presumably simply moved back to the Gulf of Cambay, from whence they had come, and continued to participate in the Red Sea trade, as Professor Das Gupta tells us, into the eighteenth century.[78]

Although we have demonstrated that vanias in the late sixteenth century overtook Muslims in controlling trade from Gujarat to the Red Sea, we have yet to show them actually voyaging themselves to the so-called "Muslim lake." We will present evidence first of vanias sailing to this area, and to the Persian Gulf, and then of them actually settled in these Muslim-dominated areas.

There is evidence from 1576 concerning "the Muslims and Hindu merchants who come to Portuguese India in their ships from the Red Sea,"[79] and another account of 1597 mentions the "baneanes," who are the main merchants trading to the Red Sea.[80] Seventeenth-century accounts describe vanias sailing in their own ships from Surat to the mouth of the Red Sea, and to the Persian Gulf; Hindu and Muslim Gujaratis sailing to the Red Sea; vanias travelling from Suakin to Surat; from Surat to Mocha; and from Gujarat to Sana.[81] As another straw in the wind, van den Broeke in 1616 noted that on the ships from Diu entering the Red Sea, "the majority of the merchants on board were Banyans, a few were Moors or Armenians . . . ."[82] There are even three instances, to my knowledge unique, of vania or Hindu nakhodas. In 1623 Sivaji Vania was nakhoda of a Surat ship sailing from Mocha with goods belonging to Dabull. At the Hadramaut port of Burum the English in 1624 talked to the nakhoda "Inderjee" (presumably Indraji and a Hindu).[83] "Lalegi Dossy", number 75 in the appendix to this chapter, is a Hindu, probably a vania. It should, however, be stressed again that we are not trying to show that all overseas trade was done by Hindus. Our aim is simply to establish that they could and did trade and travel overseas. Clearly most of the sea travellers were Muslim; most vanias financed trade, but did not travel. As one example, although we know the names of three Hindu nakhodas, we also have in the English records the names of thirteen Muslim ones,[84] while those listed in the appendix to this chapter are also nearly all Muslim.

We may finally list briefly the evidence of vanias resident in Middle Eastern ports, keeping in mind first that to get there they must have undertaken at least one sea voyage, and second that there is evidence they returned home in old age, or simply to visit.[85] We have news of vanias trading in Persia, and settled in Baghdad.[86] There is evidence of vanias, or Hindus, resident in Hurmuz in the whole period from 1400 to the seventeenth century.[87] These vanias had their own captain or captains during Portuguese times.[88] Aubin's important article notes a population of 3,000 hearths at 1600; 500 Portuguese, 300 Indian Christians, 800 "gentiles", 200 Jews, and 1200 Muslims. Among the "gentiles" were brahmins and, in the mid sixteenth century, a community of yogis. In fact, noting the use of betel, the protection of cows, and the Indian words used in maritime matters, Aubin prefers to call the culture of Hurmuz Indo-Islamic rather than Middle Eastern.[89]

Vanias also settled all along the Hadramaut coast around to Aden,[90] and within the Red Sea itself. They were found at Masua and Suakin on the Ethiopian shore, and at Aden, Mocha and Jiddah. Van den Broeke discovered over 3,000 of them at Mocha with, as in Aden, Masua and Jiddah, their own consul or Shahbandar.[91] It is clear that these vanias prospered, despite occasional harassment from the Muslim authorities and a generally second class status in these strict Muslims areas.[92] However, as a Gujarati, a Muslim in this case, told Gandhi in South Africa in 1893, "only *we* can live in a land like this, because for making money we do not mind pocketing insults . . ."[93]

snow sportsmen and the like. Equally interesting and chequered is the history of this strategic region of India. Fortunately, unlike any other part of the Indian sub-continent, it possesses an almost complete connected series of written records and chronicles of the past history of the State, namely, Kalhana's 'Rajatarangani' upto AD 1006; followed by Jonarji's chronicle upto AD 1420; followed by Pandit Srivara's writing upto AD 1489; followed by Pragya Bhatt's 'Rajavalipatak' upto AD 1586; followed by recordings of Sanskrit and Persian scholars of Mughal period as also by many English travellers.

Legend has it that Kashmir valley was originally a vast lake known as the Satisar (lake of Sati, the consort of Shiva). In it dwelt a rakshasa or a demon, Jalodhava, who frequently devastated the neighbouring regions. Once when the sage Kashyapa, the King of the Nagas, was on a pilgrimage to the North of India, his son Neile King of the Kashmiri Nagas, told him about the cruelties of Jalodhava. Kashyapa sought the help of Brahma and other gods who agreed to mount a divine campaign against the rakshasa and accordingly took up positions on the mountain peaks around the lake. Vishnu directed god Shiva, to drain the lake by piercing the mountains. When this was done and the waters drained out to the west, Vishnu and other gods fell on Jalodhava and killed him. Kashyapa then sent human beings to settle in the newly drained land where they dwelt among gods and goddesses. This, of course, is a legend, but, it is a fact established by geological research that Kashmir valley was, indeed, originally a vast lake.

The Aryans who moved across Central Asia into India are the first people who can be traced to reside in Kashmir. The first recorded King was Gonanda - a relative of the King of Magadha. Kashmiri traditions say that when

[1] There are numerous shrines along the Jhelum River, called Pandav temples, e.g. at Avantipur, Pattan and Buniyar.

Magadha was engaged in battle with Krishna on the bank of the Jamuna, Gonanda went to his relative's aid as the head of a large army and besieged Krishna in a fortress. The Kashmiris fought bravely but were eventually routed and Gonanda was killed by Krishna's brother. While no mention is made of Kashmir in the Mahabharata the Kashmiris cling to the belief that the Pandavas once ruled in Kashmir[1]. Between the pre-historic and mythological years of Kashmir history and the reign of Ashoka, Kashmir constituted a part of the Gandharvan Empire in Eastern Afghanistan. It came under Persian rule in the 5th Century B.C. However, it is only during the reign of the Great Ashoka that Kashmiri political history begins to take root.

Ashoka, the grandson of Chandragupta Maurya, was sent on a mission to the North by his father to restore order in Taxila (Now in Pakistan, a world famous university of its time, contemporary of the Nalanda University in Bihar). His acquaintance with Kashmir grew thereafter, and he is credited with having founded the original city of Srinagar. He propagated Buddhism in this region and the valley became a renowned centre for the study of Buddhism. Recently a large number of Buddha statues, pillars and other artefacts were found when permanent accommodation for the Army Signal Regiment was being constructed. These have been carefully preserved in Srinagar cantonment. Jaluka, the son of Ashoka, became independent ruler of Kashmir after the death of the Emperor. It became a centre of higher learning in rhetoric.

The Karkota dynasty came to power in the middle of the seventh century. The first king was an expansionist and extended his dominion from Poonch to Taxila. The zenith of the Karkota dynasty was reached in the mid-eighth century under the fourth ruler, Lalitaditya (724 - 760 AD). He was a good administrator and maintained a large army. He constructed the famous sun temple on the plateau about the village of Matton (Martand in Sanskrit), the magnificent ruins of which amaze, even in these times, all beholders by the unique grandeur of its architecture.

The reign of Kashmir passed into the hands of Lahoras in the beginning of the eleventh century. About that time the well known Mahmood Ghaznavi began a series of raids into the Punjab though he did not for some time attempt any incursion into Kashmir. He did make an attempt to invade Kashmir in the second decade of the eleventh century, but he failed to break through its surrounding mountain ranges.

The first Muslim ruler of Kashmir was Rinchen, an adventurer from Ladakh who embraced Islam when the Brahmins refused to accept him as a Hindu. Following his demise, there was a brief revival of Hindu rule under Kota Rani, the widow of the last Hindu King. This short-lived restoration was ended when Shah Mir, a Muslim adventurer from Swat, staged a military coup resulting in the suicide of Kota Rani. In 1343 AD, he seized the throne and declared himself king under the title of Shams-ud-Din thus founding the Sultanate dynasty which ruled Kashmir for two hundred and twenty-two years. During the period of the Sultanate, Islam became firmly established in Kashmir and its artistic and cultural expression flourished. Among these Sultans may be mentioned Sikander, called the Iconoclast (1394-1413), whose reign is remembered chiefly for his relentless persecution of Hindus and destruction of their majestic temples, the notable among these being the famous sun temple at Martand which he could not completely demolish in spite of strenuous efforts. He let loose a reign of terror against Hindus till, as the tradition goes, only eleven Brahmin families survived this holocaust by taking shelter in remote inaccessible mountain hideouts or migrating to the neighbouring Hindu principalities like Kishtwar and Rajouri. But human history now and then provides unexpected healing episodes. In Kashmir this was provided by the long reign (1422-1474) of Zain-ul-Abidin,

Sikander's son and successor. He restored a uniquely constructive and liberal rule leading to the restoration of Hindu faith, rebuilding of their vandalized temples and the recall and rehabilitation of most of those who had sought shelter in the nearby peaceful regions. Apart from this unique achievement, Zain-ul-Abidin distinguished himself as a keen promoter of learning in both Sanskrit and Persian languages. He was also a gifted and constructive genius and enriched his dominion by building canals and planting orchards of various imported fruit species. No wonder, in popular parlance, in Kashmir, he is still called "Badshah" or the Great King. Following the benign and enlightened rule of Badshah, a supplementary harmonising process may be mentioned as the evolution of the cult of Shaivaite philosophy among the Hindus and the rise of the Sufi Rishi cult among the Kashmiri Muslims. The outstanding exponent of the former trend was a wandering Hindu woman saint named Laleshwari (late fourteenth century), and of the latter Sheikh Nurdin Nurani (early fifteenth century), a Muslim Sufi saint whose mortal remains lie buried in the celebrated shrine of Chrar Village. Thus was evolved the characteristic popular ethos now usually called "Kashmiriat".

After the Shahmir Sultanate dynasty came to an end, there was a brief period when the Chaks, descendants of the Dards, ruled Kashmir. A series of internecine wars brought the administration of the state to a standstill. So desperate was this situation that a deputation of nobles went to the Moghul court in Agra and invited the Great Akbar in 1587 to invade and annex Kashmir to his dominion.

Akbar's rule was marked by a number of religious, social and economic reforms. Towards the end of Moghul rule, its hold on Kashmir became increasingly tenuous. Effective control over the vast empire gradually shrank after the death of Aurangzeb. The next dynasty to rule Kashmir was that of the Afghan, Ahmed Shah Abdali. Conditions under the Afghan Governors were far worse than under the Moghuls and for the sixty years of their rule, Kashmir was a prey not only to natural disasters, against which an ineffective and corrupt administration could offer no relief, but also to constant persecutions and oppressions. With the rise of the Sikhs in the Punjab, it was inevitable that Afghan dominance in North India should be gradually undermined. As dissensions within the Afghan territories increased, its power weakened and the strategic control of the area passed into the hands of the new strong man of India, Maharaja Ranjit Singh.

In 1809, when Ranjit Singh was expanding his rule, the Britishers employed pressure tactics and persuaded him to sign the treaty of Amritsar. This treaty deprived him of control over territories south of the Sutlej river and limited his expansions to the North. In 1818, he succeeded in capturing the territories held by Nawab Mazaffar Khan of Multan. In the following year, he annexed the province of Kashmir wresting it out of the hands of the Afghans.

Ranjit Singh occupied this province in 1819 and appointed Diwan Chand, commander of the forces, as the first Governor of this province. The second important Governor was Diwan Moti Ram. He was kind-hearted and liked by the people. By his just and humane conduct, he restored confidence in the Valley. It was during his governorship that Moorcroft arrived in Kashmir from Ladakh where he had gone by the Kulu route two years earlier, ostensibly to find out ways and means for extending British commerce in Central Asia, but really to collect geographic, political and military information. It was the first indirect incursion of a Britisher into Kashmir.

Kashmir remained under Sikh rule till 1846. The Sikhs who succeeded the Afghans were not so cruel, but were hard and exacting masters. Both Hindus and Muslims considered their rule far better and far more humane than that of the Afghans. The standard of living of the masses was, however, very low.

Gulab Singh, the man credited with the founding of Jammu and Kashmir State was a highly controversial figure in his time and is mentioned by historians as unscrupulous, though dedicated to his subjects. He was soft-spoken, ambitious, far-sighted and self-seeking. It was he who provided peaceful and orderly administration to the State. In 1846, after the death of Maharaja Ranjit Singh, he disassociated himself from the Sikh Kingdom, struck a deal with the British and was made the ruler of Jammu and Kashmir, for a consideration. In addition to the paltry sum of money he paid to the British, he also ceded to them the district of Hazara where, at Abbottabad, they established the home of the Punjab Frontier Force to control the North West of India.

Gulab Singh was born in 1792 and early in his childhood was sent to live with his grandfather, a stern old warrior who gave him a thorough training in all martial arts. At the age of sixteen, he took part in the defence of Jammu when Ranjit Singh's general Hukum Singh had tried to annex this State. Later, Gulab Singh was credited with quick and courageous efforts to quell the rebellions and defend a fort while serving under Diwan Khushwant Rai.

Maharaja Ranjit Singh was informed about these brave actions of Gulab Singh and later sent for him. He joined the forces of Ranjit Singh and took part in several campaigns against the Afghans in which he distinguished himself as a military commander and fearless warrior. In the ill-fated expedition of 1814 against Fateh Khan, the Afghan Governor of Kashmir, Gulab Singh's regiment played a prominent part. While Sikhs were retreating; he reassembled his forces and brought them back safely. At the siege of Multan, Gulab Singh's personal bravery in recovering the dead body of Ranjit's favourite sirdar in the thick of fighting won him the praise and admiration of his master.

In 1821, Gulab Singh effected the conquest of Kishtwar. He punished the ruler of Rajouri. All these distinguished services to the Lahore Court were rewarded by the Maharaja with the grant to Gulab Singh and his successors the principality of Jammu with the hereditary title of Raja.

Maharaja Ranjit Singh died in 1839, thereafter the grip of Sikh rule became loose. Gulab Singh thought of expanding his rule by annexing Ladakh and Baltistan. General Zorawar Singh was entrusted with this formidable task.

Marching through Kishtwar, the Dogras fought a battle in Pushkyum valley where the Ladakhis were defeated. Ladakhis again marched down to attack the Dogras near Langkartze in the Suru valley, but on arrival of Zorawar's forces they fled. They were pursued upto Leh where King Gyalpo had to surrender and pay an indemnity of Rs.50,000 and a yearly tribute of Rs.20,000 to the Dogra chief. He also conquered Baltistan and Gilgit without much difficulty.

In 1841, Zorawar Singh with a force of 5,000 troops marched towards Lhasa to conquer the whole of Tibet. Having overrun the two provinces of Rudok and Gartok, he proceeded 18 kilometres inside the Tibetan territory. These two provinces which produce Pashm, the finest wool of Tibet, also contained the wealthiest and most sacred of its monasteries held by the Buddhist ruler of Tibet as the most valued possessions. Thus, Zorawar Singh provoked the wrath of Lhasa. On 12th December, 1841 when the Dogras of the hot plains and low hills of the Punjab and Jammu were losing their hands and feet from frost bite, the Tibetan forces under General Shastra delivered an assault, mercilessly killing their enemy. Zorawar Singh was killed and only 25 soldiers of his army returned to Kashmir.

Gulab Singh did not accept this defeat as final. Next year, he again sent Diwan Hari Chand Wazir to re-conquer Ladakh. However, the intrigue of Lhasa to foment trouble for the Dogras continued upto as late as 1891, and the Maharaja of Jammu and Kashmir had to keep a strict watch over the movement of Lamas across the various passes. With the conquest of Ladakh and Baltistan, Gulab Singh's dominions expanded. He is the only ruler to have extended the frontiers of India beyond the Himalayas. In 1846, Gulab Singh acquired the territory of Gilgit. However, in 1889 a part of the territory called Gilgit Agency bordering Afghanistan and Russia was taken over by the British who apprehended Russian expansionist tendencies. The Maharaja, of course, retained sovereignty over the Gilgit Agency and also complete control over the balance of the territory called Gilgit Wazarat.

Maharaja Gulab Singh installed his son Ranbir Singh on the gaddi in 1856 when his health began to fail. Ranbir Singh was born in 1829 A.D. and passed his youth in the company of his father and took part in several of his campaigns, and thus he was trained as a proficient soldier and efficient commander. He was a good administrator also. In 1885, Maharaja Pratap Singh succeeded Maharaja Ranbir Singh.

Maharaja Pratap Singh died leaving no male heir and in 1925 the gaddi of Jammu and Kashmir passed on to his nephew, Raja Hari Singh. Born in 1895, the young prince had received his education at Mayo College, Ajmer.

Maharaja Hari Singh brought many changes in his administration like 'Agriculturists Relief Regulation' to relieve farmers from the stranglehold of money-lenders; introduction of compulsory primary education and preference to the State subjects for appointments in government services. He tried to persuade the British to abolish the Gilgit Agency and hand over its control to the Maharaja. Gilgit Agency was originally separated from

Gilgit Wazarat in 1889 as the British wanted to keep a direct eye on the Russian designs. Maharaja Hari Singh failed to make any progress in his demands. In 1935, he finally was forced to withdraw his own administrative machinery from the Wazarat portion of Gilgit and hand it over along with the Gilgit Agency to the British on a sixty years' lease.

Poverty, very high taxes, trade depression and widespread unemployment made the people of the state discontented. Thus, Kashmiri Muslims, who had studied outside the State, gave expression to the people's discontent. Mahatma Gandhi's non-cooperation movement in India set an example for conducting a campaign against unjust rule. The British also encouraged the agitators in Kashmir with material support because they had doubts about the loyalty of the Maharaja towards the British rule.

The police firing on 13th July, 1931, on a crowd of four to five thousand Muslims who staged violent demonstrations outside the court killed 21 Muslims. The Muslim population became highly incensed and proceeded to wreak vengeance on local Hindus. The communal rioting spread like wild fire both in Kashmir and Jammu. The Maharaja appointed a committee under Sir Barjor Dalal to enquire into the cause of the riots, but it was boycotted by the Muslims. At this critical stage, the British insisted that the Maharaja should be more responsive to the needs of his Muslim populace. This pressure did have some effect and things improved.

Communal agitation subsided when the Jammu and Kashmir Muslim Conference was organised in October 1932 with Sheikh Abdullah as its first President. In the next two years, the need for forming a united organisation was loudly proclaimed by leaders of all communities to seek freedom from the prevailing autocratic rule of the Maharaja. In June 1938 the Working Committee of the

Muslim Conference passed a resolution and opened the party doors to all irrespective of caste, creed and region. In 1939 it transformed itself into the National Conference. It was this party which played a decisive role at the time of partition in 1947.

The National Conference gave moral and political support to the Quit India Movement in 1942 launched by the Indian National Congress. In 1944, the National Conference passed a resolution to establish in Jammu and Kashmir a socialistic pattern of society and a democratic government. The Maharaja conceded a few demands and agreed to induct two popular Ministers in the Cabinet. But Mirza M.A. Beg, the National Conference nominee resigned after two years due to administrative non-cooperation. The National Conference then started the 'Quit Kashmir' Movement in 1946 and the State Government let loose a reign of terror and arrested leaders of the National Conference. At this stage, national leaders supported the move of the National Conference and Jawaharlal Nehru visited the State. Mahatma Gandhi also paid a visit to the valley in August 1947. The Maharaja dismissed Ram Chander Kak from Prime Ministership and appointed General Janak Singh in his place. Kak had started his career as a librarian and rose to become the Prime Minister. He was married to an English woman and was well up with the British establishment. During the fateful period of 1946-47, the Maharaja suspected his loyalty which resulted in his dismissal. There was an outburst of spontaneous joy among the people of all communities and they looked for a clear lead from the National Conference with regard to the most important question of accession of the State to India or to Pakistan.

On 17 June, 1947, the Indian Independence Act was passed stating that on 15 August, 1947, the British would relinquish their authority in India which was to become an independent country. On the lapse of paramountcy of the

British crown over all the princely states, of which there were 562 at the time of independence, the problem arose as to what was to be done with them, because they were not directly ruled by the Government of India. To this effect, the British made an announcement on 3 June, 1947, simultaneously with the announcement of the Partition scheme :-

> "His Majesty's Government wishes to make it clear that the decision announced above (about partition) relates only to British India and that their policy towards the Indian States contained in Cabinet Mission Memorandum of 12. 1946 remains quite unchanged."

The legal provision under which the princely states could enter into a federal relationship with the successor Government or Governments in 'British India' was to be found in the Indian Independence Act of 1947 and the Government of India Act of 1935. These two Acts of the British Parliament, which created the legal basis for independence of India provided that a State could accede to the Dominion of India or Pakistan by an Instrument of Accession executed by the ruler. It was also provided that the Indian States acceding in this manner shall become an integral part of India or Pakistan.

But, to prevent a dislocation of the arrangements with regard to customs, transit and communication, posts and telegraphs or other like matters during a period - between 17 June, when the Government of India Act was passed and 15 August, when India became independent - the ruler of a princely state could enter into a standstill agreement with either or both the independent dominions.

The Maharaja of Jammu and Kashmir delayed his decision in regard to accession as he had been dreaming of an independent Jammu and Kashmir State. Fearing a breakdown of the communication system through Pakistan and the rich export trade with India, the Maharaja sought from both dominions a standstill agreement to come into effect on 15 August, 1947.

The history of the standstill agreement is contained in the telegrams exchanged between the Prime Minister of Jammu and Kashmir, on the one hand, and the Prime Minister of Pakistan on the other hand. There was a similar series of telegrams between the responsible Minister in New Delhi and Prime Minister of Jammu and Kashmir. A standstill agreement between Pakistan and Jammu and Kashmir was arrived at with regard to her (State's) communications, supplies, and post office and telegraphic arrangements.

On the same date (12 August, 1947), an identical telegram was sent to the Government of India stating that the Jammu and Kashmir Government would "welcome standstill agreement with Union of India on all matters on which these exist at the present moment, with the outgoing British Indian Government. It is suggested that the existing arrangement should continue pending settlement of details and formal execution of fresh agreements.[1]" These arrangements, according to the treaty of Amritsar of 1846, comprised the use of the Indian forces if there was internal rebellion or invasion by a foreign government.

The Government of India did not accept the offer at once and telegraphed back :

"Government of India would be glad if you or some other Minister duly authorised in this behalf could fly to Delhi for negotiating standstill agreement between Kashmir Government and Indian Dominion. Early action desirable to maintain existing arrangements and administrative arrangements.[2]"

But, before the Prime Minister of Jammu and Kashmir could go to Delhi, the State was invaded by Pakistan. Therefore, the standstill arrangement which the State was trying to arrive at with India was interrupted and other developments followed.

The partition of India in 1947 has become a reality and is to be accepted as a fact of history. However, if we understand the circumstances which culminated in the partition of the country, we will be able to appreciate the role played by the Saviours of Ladakh with a degree of detachment, without in any way, diminishing our admiration for the heroic role played by them. It is necessary to do so because the world is moving to a new age - the age of unity of man, global economy and eventually earth citizenship.

The seed of partition of India was sown after the uprising of the Bengal Army in 1857 - the British call it the Indian Mutiny. It was an uprising by all Indians against the British rule. After the uprising was ruthlessly put down, a conscious and deliberate decision was taken by the British to follow a policy based on the traditional principle of 'divide-et-impress' meaning 'divide and rule'. The policy runs as a continuous thread throughout the period of British rule in India especially during their final ninety years or so. The first impact of this policy was on the Indian Army, which was restructured on the basis of units and sub-units being composed of ethnic, caste and religious groups, so that one could be used against the other if ever there was a need. Its final impact was on the political life in the country when Hindus and Sikhs versus Muslims were pitched against each other with intense ferocity. However, this fact should not be viewed as an evil invented by the British. Divide and rule is an ancient technique of imperial governance. Since independence, the national rulers of India and Pakistan have also used it with immoral double-speak and cunningness in each country.

PARTITION OF INDIA

A British experience which contributed to the partition of India was what they did in the Middle East. After World War II, the Arabs were liberated from the Ottoman Empire. At that stage, the British adopted a policy to control the 'liquid gold' - oil - of the Middle East. Based on the clever advice of two well known Arab experts, a two-pronged policy was adopted. One expert was the famous historian Arnold Toynbee and the other Philby (senior), Harry St John Bridge, the famous Arabist and the father of Philby (junior), Kim who defected to the USSR from the Middle East, and ended up as a general in the KGB. The two facets of this policy were :

- To divide the Arabs into numerous large and small States and Sheikhdoms, and
- To promote Islamic fundamentalism so that Arabs would not modernise and assert their control over their most valuable mineral wealth. It is well known that fundamentalist leaders like Ayatollah Khomeini of Iran, the Grand Mufti of Jerusalem and many activists of Muslim Brotherhood Party were funded at one stage or other by the British intelligence (MI 5).

The above policy paid handsome dividends to the British. Britain and America continue to exercise control over oil, and much of the oil wealth is invested in the Western economies.

Control over the Middle East was exercised by the British from India. As Indian independence approached nearer, their experience of the Middle East prompted them that it would be easy to manipulate a small and economically weak state like the proposed Pakistan than a monolithic united India. It is precisely for this reason that the date for granting independence to India was hurriedly advanced from 1948 to 1947, ostensibly as a concern for the Indian political opinion, as also the serious Hindu/Sikh versus Muslim riots that broke out in the country. In reality, it was based on the British intelligence report that Mohd Ali Jinnah was suffering from acute T.B. (a closely guarded secret) and may not live till 1948 (he died in September 1948). The British were concerned that, in the absence of Jinnah other Muslim leaders may well accept a united India. It is well known that Mahatma Gandhi had tried and failed to hand over the Prime Ministership of united India to Jinnah. Pandit Nehru and Sardar Patel did not accept Gandhi's advice. The British, knowing the tremendous hold of Gandhi on the Indian people were apprehensive that Gandhi may yet prevail after Jinnah was gone.

BLUEPRINT FOR PARTITION BY BHAI PARMANAND CHIBBER

It is of interest to note that the first thought of dividing India on communal lines was put forward by Bhai Parmanand Chibber in a small booklet published in 1923[3]. This was followed in 1924 by a few articles by Lala Rajpat Rai, in the 'The Tribune' newspaper of the Punjab. His thrust, in his own words was :

> "Punjab should be partitioned into two provinces, Western Punjab with a large Muslim population will be a Muslim governed province and the same principle can apply to Bengal. Under my scheme the Muslims will have four provinces, the NWFP, West Punjab, Sind and Eastern Bengal. But it should be distinctly understood that it is not a united India. It means a clear partition of India into a Muslim India, and a non- Muslim India[4]."

It was much later that some Muslim thinkers put forward the concept of the two-nation theory. They demanded Pakistan for the benefit of the Muslim community of India. The partition of the country finally took place as per the Bhai Parmanand plan which was also the heart of the Rajagopalachari formula put forward by him in 1944.

The 'direct action' launched by the Muslim League in August 1946, resulting in large scale communal rioting, which was the final straw. Jinnah's call to 'divide and quit' was music to the British ears and so the division of the country became a reality.

The concept of dividing the country on communal lines is being questioned by many youth in all the three countries: India, Pakistan and Bangladesh. And, this is despite a stranglehold of the establishment in Pakistan which brought about this tragedy on the community. Some well-meaning thinkers in Pakistan do put forward the argument that when brothers in a joint family cannot live together, then, it is best for them to part and live separately. However, they quickly retrace their argument when they realise that the largest number of Muslims did not separate out from their geographic, cultural and ancient roots and stayed on in India.

Some young Muslim thinkers have tried to analyse the partition. In a well researched book 'Sole Spokesman[5]' - a Ph.D. Dissertation at Cambridge University in Britain, Ayesha Jalal of Pakistan concluded that Mr.Jinnah's demand for Pakistan was essentially a pressure tactic to win equality for the Muslims with the majority community in sharing power in a united India. However, the compulsion of events precipitated by severe communal rioting led to partition. There is some merit in this line of thought. It was essentially this obsession with equality between unequal numbers that prevented Pakistan from adopting a democratic arrangement to let East Pakistan with a large population have a proper role in running the country. And later, when, in 1971, the general election gave East Pakistan a decisive majority, the opportunity to rule the country was denied to them. Thus an obsession for equality between unequals eventually led even to the break-up of Pakistan and the birth of Bangladesh. Again, it was the Muslim community that suffered the most. Thus the partition of India, in event, has turned out to be a grave injustice and tragedy for the Muslim community because it has got divided into three nation states and :

- The largest number of Muslims continue to be in India. They have been suspected, exploited as vote banks, have developed a ghetto mentality and have been manipulated by the radicals in their own community. Economic benefits and educational opportunities of a democratic country have not reached them in the same measure as for others in India. It is relevant to note that the Muslim population of India at 126 millions make them the second largest Islamic populace in the world after Indonesia.
- The second segment is in Bangladesh which is among the poorest and the most densely populated countries in the world.
- The smallest segment is in Pakistan where it has been directly or indirectly ruled by the military for almost four decades. Democracy and development have failed to take firm roots there. The Urdu speaking high caste Muslims who migrated from India to the promised homeland of Muslims are having a tough time in Sindh and some are languishing in refugee camps in Bangladesh because Pakistan does not accept them.

PRINCELY STATES IN INDIA

In the Independence Act for India, passed by the British Parliament, the future of 562 micro, mini and large princely states in India was left completely in the hands of their rulers. They could join India or Pakistan or stay independent. Many analysts feel that the formulation of this provision had a sinister purpose. It was hoped that like the Arab world, India would also be fragmented into numerous nation states to enable the British to perpetuate their influence in South Asia. However, the political movement in the country had developed very strong urge for democratic rule in the princely states also. As a result, a vast majority of the rulers acceded to India or Pakistan in deference to the wishes of their people, of course advised or persuaded by the political leaders in India and Pakistan. However, three States faced a dilemma due to the handling of the situation by their rulers. These were :

a) **Junagarh:** A micro State of 0.6 million people ruled by a Muslim Nawab, while the population was predominantly non-Muslim. When the ruler acceded to Pakistan, people rose in revolt, the ruler fled to Pakistan and the people invited India to take over the State which was done after holding a referendum in February 1948.

b) **Jammu and Kashmir:** Geographically a very large state of India, but sparsely populated. A population of about 4.5 million of which seventy-five per cent were Muslims, but unevenly spread. While the Kashmir Valley, Baltistan and Gilgit were predominantly Muslim, in Ladakh and Jammu region the non-Muslims had a slight majority. The Hindu ruler entered into a standstill agreement with Pakistan and was negotiating for a similar arrangement with India, when the Tribal Invasion from Pakistan was launched on 20 October 1947 - about two months after India's independence on 15 August 1947. The invasion forced the Maharaja to accede to India.

c) **Hyderabad:** Located in South India, it had the largest population of all the princely states. Of the 25 million people the non-Muslims were a predominant majority, but the ruler, the Nizam of Hyderabad was a Muslim. Like his counterpart in J&K he delayed a decision till September 1948. The fundamentalist Islamic group of Razakars led by Kasim Rizvi prevented the Nizam from acceding to India. The Indians mounted a police action in June 1948 and the Nizam eventually had to accede to India.

There is a psychological hurt in the minds of the Pakistani establishment. They feel that Junagarh and Hyderabad which were predominantly non-Muslim States but had a Muslim ruler were coerced by India to accede to India. Consequently, J&K state which was predominantly inhabited by Muslims should have acceded to Pakistan.

THE TRUTH ABOUT ACCESSION OF KASHMIR

As discussed earlier, the rulers of the 562 princely states in India had the sole right to accede to India or Pakistan, or to stay independent. The rulers of the two major States, namely J&K and Hyderabad delayed making a decision beyond 15 August 1947, when the British rule ended in India. The ruler of J&K, like the Nizam of Hyderabad, seriously toyed with the idea of independence, and hoped that standstill agreement would give him time.

It is appropriate at this stage that the basic truth about the preparation and launching of the Tribal Invasion is recorded to appreciate the fateful comment in the book 'Memoirs of Jinnah'[6] that this *invasion was a criminal folly which sealed the fate of Kashmir*; meaning that it left no alternative to the Maharaja except to accede to India. It appears that Jinnah was informed of the Tribal Invasion only a few days before it was to be launched and invited to come up from Karachi to Abbotabad for a triumphant entry into Srinagar. Jinnah was aware that both the Maharaja and his Prime Minister Mr. Ram Chander Kak, of J&K , hated Nehru bitterly. He had sent his personal secretary Khurshid to Srinagar to facilitate the Maharaja's accession to Pakistan. In event, Khurshid got trapped in Srinagar by the Tribal Invasion and was repatriated only in 1949 after the ceasefire.

From 15 August 1947 (when India and Pakistan became independent) to 20 October when the Tribal Lashkar entered Jammu and Kashmir it was only about nine weeks in which this 'folly' was prepared and committed. It is well to appreciate that there was no coherent central government in Pakistan. Delhi was the seat of central government for the whole of British India till 14 August. Overnight the whole thing changed. Karachi was improvised as the capital of Pakistan where the Constituent Assembly of that country was convened. Components of the Composite Indian Army were still undergoing a sort of surgery to move them from their homes and hearths to the new nations. An example will give the reader a flavour of the conditions.

The second and third post-war regular course cadets were being trained at the Indian Military Academy, Dehradun, as officers for the undivided Indian Army. In the month of July they were told that the Indian Army was to be divided between the two countries. Non-Muslim cadets who belonged to the areas included in the proposed State of Pakistan could opt either for Pakistan or India, but Muslims from that region had compulsorily to go to Pakistan. Similarly, Muslim cadets who belonged to the residual Indian region could opt either for India or Pakistan, but no Muslim cadets from the proposed Pakistan could stay in India. In response to these options, many non-Muslim cadets including the author, who belonged to West Punjab or NWFP, opted for Pakistan. There was no inkling at that stage that the biggest migration of humanity in history was in the offing. It was only when bloody riots broke out in August 1947 that the chasm became real and original choices were revised. Cadets going to Pakistan were moved from Dehradun in September 1947 at virtually six hours' notice. Similar surgical operation was going on all over the country.

In this fluid and emotionally surcharged situation many individuals in Pakistan vigorously acted to ensure that Kashmir somehow became a part of Pakistan.

The more prominent among these was Mr Liaqat Ali Khan of UP who became the Prime Minister as well as the Defence Minister of Pakistan.

Another was Khan Abdul Qayum Khan was took over the Chief Ministership of North West Frontier Province from Dr Khan Saheb who having won the election in 1946 as a Congressman was against the partition of the country. Dr.Khan Saheb resigned as Chief Minister after a referendum was held to decide if NWFP should join Pakistan or India (the Congress party boycotted the referendum because the Muslim League used the slogan for the people to choose between Quran or Geeta).

Yet another was Lt.Col.Iskander Mirza who later became the President of Pakistan. He was an officer in the political department in the North West Frontier Province dealing with the Tribal affairs. On independence he became the Defence Secretary of Pakistan. The GHQ of Pakistan was hurriedly being organised at Rawalpindi in the former headquarters of the Northern Command.

Khan Abdul Ishaq Khan, later to become the President of Pakistan, was in it at a junior level. He was a competent officer of the provincial civil service of the North West Frontier Province. In 1947 he was personal secretary to Dr Khan Saheb of the Congress party and continued to remain close to the levers of power when Khan Abdul Qayum Khan took over as the Chief Minister (Ishaq exhibited this remarkable capability throughout his career, eventually becoming the President of the country).

Lt. Col. Roger Bacon, the British Political Agent governing the Gilgit area which was on 60 years lease to the British Government as agreed to by the Jammu and

Kashmir State, also played a role. The lease was terminated on 3 June 1947, on the eve of independence. Bacon left Gilgit after handing over charge to Brig. Ghansara Singh of the Jammu and Kashmir State Forces who flew to Gilgit on 1 August to become the Governor of Gilgit. Bacon went to Peshawar and took over as the Political Agent of Khyber region, the hub of all the Tribal affairs. (His role in Gilgit will be discussed later).

In non-public discussions with Pakistani scholars who have sifted the *truth* about the Kashmir problem, they blame Khan Abdul Qayum Khan, the Chief Minister of North West Frontier Province, as the key figure in organising this invasion. He was of Kashmiri origin and obviously had as emotional an attachment to Kashmir as Nehru, also a Kashmiri. In the Frontier Scouts, Qayum had a readymade set-up to help in this folly. He was ably assisted by Lt. Col. Iskander Mirza who became the Defence Secretary and knew how to handle the Pathans.

Frontier Scouts were, and continue to be recruited, from various Pathan tribes like Mehsuds, Mohmads, Khataks, Wazirs, Tochis and so on. It was, and still is, a very effective paramilitary force officered by volunteer British officers of the then Indian Army (now Pakistan Army Officers). Since service with Scouts provided combat experience and much money, officers competed for the few vacancies that were available and some of the best officers went to the Scouts. The force was used to maintain law and order in the autonomous Tribal belt between British India and the Afghan border. It was and still is a private army under the control of the government of NWFP. Consequently, for Abdul Qayum Khan, it was quite a simple matter to organise the Tribal Invasion.

There is an interesting episode on how the Gilgit Scouts, in coordination with the Tribal Invasion, played a direct role in Kashmir affairs. As we discussed earlier, the

British ended the lease of Gilgit on 3 June 1947 and Brigadier Ghansara Singh of the State Army took over as the Governor. At this stage, the Gilgit Scouts, a wing of the Frontier Scouts, controlled from Peshawar, also reverted to the control of the J&K State. This wing was raised during the lease period from among the local inhabitants to keep a watch on the border with Afghanistan, Russia and China. On the termination of the British lease, the J&K State had to replace the Commandant of the Scouts. Major General M. Scott, the Chief of J&K State Forces, was anxious to get a British officer who was familiar with the Scout culture. The choice fell on a young 24 years old Scott, Major William Brown, who had already done a three-year stint with the Gilgit Scouts and had also earlier served with the Tochi and Chitral Scouts. He was appointed as none older was available. Brown selected one Capt. A.S. Mathieson as his second-in-command. Incidentally, Brown was awarded one of the highest Pakistan awards in 1993 for his role in Gilgit about 46 years earlier.

As the Commandant of the Scouts, Brown started reporting to Brigadier Ghansara Singh, the Governor, but it is on record that "he was emotionally committed to the cause of Pakistan." [7] Along with his second-in-command they undertook two steps. They put down the 'Gilgit Liberation Front' which was preparing to declare independence, and had infiltrated even into the ranks of the Scouts. And, secondly, they decided that should the Maharaja accede to India they would stage a coup d'etat and pass on Gilgit to Pakistan. The code signal for the coup d'etat was fixed as "Datta Khel."[8]

The Tribal Invasion code named "Operation Gulmarg," was approaching Srinagar on 26 October when the Maharaja acceded to India and on October 27, the first two companies of Sikh Regiment were flown to Srinagar. When the Dakotas carrying them were coming in to land, the pilots were not quite sure if the airport was still in the hands of the State Forces, or had fallen to the invaders during the night. In case the airfield had been captured by the invaders, their orders were to fly back and land at Jammu. The Sikhs were then to move by road to Srinagar as far forward as possible. However, the Tribal Invaders had wasted three precious days in loot, rape and arson at Baramula, and were thus delayed in reaching Srinagar, and the city was saved just in time.

In Gilgit, Maj. Brown issued the code word "Datta Khel" on 31 October and a coup d'etat was staged. Brigadier Ghansara Singh was taken into custody by the Gilgit Scouts to be later repatriated to India through Pakistan. The most interesting part of this episode was the two telegrams[9] which Major Brown sent to Peshawar on 1 November 1947. These were :

To	:	Khan Abdul Qayum Khan, Prime Minister NWFP
From	:	Major Brown
Date	:	1st November 1947

> "Revolution - night 31 October to 1 November in Gilgit Province. Entire pro-Pakistan populace have overthrown Dogra regime. Owing imminent chaos and bloodshed Scouts and Muslim officers of State Forces have taken over law and order. Scouts and Muslim officers of State Forces running administration provisionally. Request higher authority be appointed for orders immediately, and reply through wireless – Commandant Scouts."[10]

The second telegram was sent to Lt. Col. Roger Bacon, who was the British political agent in Gilgit till 1 August, when he handed over the responsibility of governance of Gilgit Agency to Brigadier Ghansara Singh. He continued to stay at Gilgit for a few days after handing over charge, and had long meetings with Brown. On leaving Gilgit he was appointed the political agent at Khyber. It is apparent from the wording of the telegram, that Bacon was quite obviously the author of the planned coup d'etat. The telegram reads :

To : Lt. Col. Roger Bacon

From : Brown

Date : 1 November 1947

> "Coup d'etat in favour of Pakistan in Gilgit can carry on meantime, but can you help."[11]

We should appreciate the reason for the British officers of the political department playing a significant role in the Tribal Invasion in general and the coup d'etat in Gilgit. During the first half of the twentieth century, British Empire had become the largest empire in human history. 'Sun never sets on the British Empire' was a favourite tribute frequently paid to the pre-eminent power in the world. During the same period a major rival in the world was Czarist Russia and after the communist revolution in that country in 1918, the USSR. It was to arrest the expansion of the Russians through Central Asia to the South that Gilgit acquired a special geo-strategic importance. It has a central position facing Northern Afghanistan, USSR, and Sinkiang region of China. That is why Gilgit was taken on lease by the Government of India on 1 August 1935. The lease was terminated once it became clear that India would be partitioned. In the British scheme of things the creation of Pakistan was essential to have a small manipulatable and friendly country to protect their oil interests in the Persian Gulf on the one hand and on the other to prevent Soviet expansion southwards to reach warm water ports of the Arabian Sea. Karachi was considered essential for the defence of the Persian Gulf. Tne creation of Pakistan was a geo-strategic necessity for the British control of oil wealth of the Gulf because, as Sir Olaf Caroe, ICS observed later in his book, 'The Wells of Power,' "the Persian Gulf opens directly at Karachi, in a real sense it's terminus."[12] The Governor of North West Frontier, Sir George Cunnigham, Sir Frances Mudie of Punjab and other British officers of the political department were working for the 'King and Country' to retain as much leverage as possible in South Asia. Lt. Col. Roger Bacon's plan to seize the former Gilgit Agency for Pakistan was for these very considerations.

Most of the people in Pakistan are innocent of the truth about the decisive damage done by the Tribal Invasion to any possibility of Jammu and Kashmir acceding to Pakistan and precipitating its accession to India. Some of the organisers of the Tribal Invasion who played a leading role have tried to cover up this folly, by promoting the belief that India's intervention in Kashmir was the result of :

a) A conspiracy by Mountbatten who was greatly incensed by the refusal of Pakistan to accept him as the common Governor General of India and Pakistan. He is alleged to have conspired with Sir Cyril Radcliffe to award the Muslim majority district of Gurdaspur to India, and give it a land route to Jammu and Kashmir.

b) Indians fraudulently conspired to force the ruler to sign the instrument of accession when the Muslim population of Kashmir revolted against him and that the tribesmen had merely rushed to Kashmir to assist their Muslim brothers being butchered by the Maharaja's forces. Both these myths bear analysis.

GURDASPUR

There was much anxiety before the boundary award was announced by Sir Cyril Radcliffe, the Chairman of the Boundary Commission, set up to decide on the line of demarcation for the division of Punjab and Bengal. The Congress and Akali leaders were certain that Lahore would come to India, as they had given the Commission convincing evidence that most of Lahore was owned by non-Muslims. Similarly, the Muslim League leaders were sure that Gurdaspur would come to Pakistan, as the majority of the population (51.14%) in the district were Muslims. Neither side got what they had pleaded for. The Indian and Pakistani members of the Boundary Commission stuck to their stands, so Lord Radcliffe told the Commission: "Gentlemen, you have disagreed and therefore the duty falls on me to give the award which I will do later on." He gave his award on 15 August 1947. Lahore was awarded to Pakistan. This had nothing to do with Gurdaspur. Radcliffe while discussing the claim of Indian members to Lahore had observed: "How can you have both Calcutta and Lahore"[13] Gurdaspur was awarded to India because without it there would be no land route to the State of Chamba, district of Dharamsala and to Kulu, Manali, Lahaul and Spiti. Further, the upper Bari Doab Canal ran through Gurdaspur to irrigate the districts of Amritsar and Ferozepur, besides being the only source of water for the sarovar (lake) surrounding the holy Golden Temple of the Sikhs at Amritsar. This suspicion of a conspiracy by Mountbatten or Radcliffe has been decisively discounted in the seminal work about the partition of India edited by C. H. Philips of the London University[14].

INSTRUMENT OF ACCESSION

The main argument in support of the myth that accession was also a conspiracy and fraud is that the Muslims of Kashmir rose in revolt against the Maharaja of Kashmir (like what had happened in Junagarh) and the Maharaja ran away to Jammu. The Indians sent their Army to crush the revolt, while doing so, they either forced the Maharaja to sign the instrument of accession or forged it. A book written by one Alastair Lamb, titled 'Kashmir, Disputed Legacy' is cited to support this theory. It now transpires that Lamb was funded by a pro-Pakistan Kashmiri Organisation in the UK to do this book. Unfortunately, the people of Pakistan have been victims of so many internal conspiracies and coups that such theories sell very easily. The truth about accession of the State to India needs to be recorded factually.

PREPARATION AND LAUNCHING OF TRIBAL INVASION

The actuality about preparing and launching the tribal invasion has come out in many writings of individuals in Pakistan who were directly or indirectly involved in the invasion. The more relevant ones are the memoirs of the commander of the raiders. Major General Akbar Khan who adopted the pseudonym 'General Tariq' (after the Arab General who led the invaders into Spain); the book 'Raiders in Kashmir' by Maj. Gen. Mohammed Akbar Khan, Pakistan Publishers, Karachi, 1970, memoirs of General Mohammed Musa (who became the C-in-C of Pakistan) in the book 'Jawan to General,' ABV Publishers, New Delhi ,1985, 'Memoirs of Lieut General Gul Hassan' (who became C-in-C of Pakistan after 1971 war), Oxford Publishers, 1994.

Maj. Gen. Mohammed Akbar Khan won a DSO in World War II. As a Colonel, in 1947 he was the Director of Weapons and Equipment in the Pakistan Army General Headquarters at Rawalpindi. He was interviewed about his memoirs and his role in Kashmir by Brig.(retd.) A.R. Siddiqi and the account was published in Defence Journal, Karachi, June-July 1985. It is best to quote Akbar :

> *"A few weeks after partition, I was asked by Mian Iftikharuddin on behalf of Liaquat Ali Khan (Prime Minister and Defence Minister of Pakistan) to prepare a plan of action for Kashmir. I found that the Army was holding 4,000 rifles for the civil police. If these could be given to the locals, an armed uprising in Kashmir could be organised at suitable places; I wrote a plan on this basis and gave it to Mian Iftikharudij. I was called to a meeting with Liaquat Ali Khan at Lahore where the plan was adopted, responsibilities allotted and*

*orders issued. Everything was to be kept secret from the Army."** (See page 69.)

"In September 4,000 rifles were issued at various places and the first shots were exchanged with the Maharaja's troops and the movement gathered weight."

"On the 24th of October a tribal lashkar attacked Muzzaffarabad and successfully captured it. The next day they advanced and captured Uri. On the 26th they occupied Baramula. On the 27th the Maharaja fled from Srinagar and acceded to India. That evening Liaquat Ali Khan held a meeting at Lahore to which I was invited. This was to consider what action to take in view of the expected Indian intervention in Kashmir. I proposed that a tribal lashkar should attack Jammu as this was the focal point through which Indian troops would be going to Kashmir.

"The performance of the tribal lashkars had been excellent where the ground was suitable for their sniping and hit-and-run tactics... It is not correct to say that they broke their ranks and went for loot just when they were within sight of Srinagar.

"It was part of their agreement with Major Khurshid Anwar of the Muslim League National Guards who was their leader that they would loot non-Muslims. They had no other remuneration. Major Khurshid Anwar had been an emergency commissioned officer in World War II in the Supply Corps of the Indian Army. Then he joined the Muslim League and he had been appointed commander of the Muslim League National Guards. In September, 1947 when the Prime Minister launched the movement of the Kashmir struggle, Khurshid Anwar was appointed commander of the Northern Sector. Khurshid Anwar then went to Peshawar and with the apparent help of Khan Qayyum Khan raised the lashkar which assembled at Abbottabad and with which he entered Muzzafarabad on the 24th of October 1947 - reached Baramulla where he delayed the lashkars for two days for some unknown reason.

"Khan Abdul Qayyum Khan had apparently helped Major Khurshid Anwar with the raising of the lashkars on the frontier. Thereafter he continued to take active interest in

Kashmir and helped with the tribal lashkars throng the Kashmir operations."

Maharaja Hari Singh intensely disliked Jawaharlal Nehru and other Indian Congress leaders who were, for long, supporting Sheikh Abdullah and his political party - the National Conference - which was agitating against the Dogra rule. In 1946 when Nehru tried to go to Srinagar in support of National Conference movement, he was detained at Uri and then expelled from the State. He had locked up the Sheikh and a few other leaders for a while during the fateful summer of 1947. His dream was to make Jammu and Kashmir an independent country as the Switzerland of Asia. When he learnt that his Prime Minister Ram Chandra Kak with his English wife had promised Pakistani leaders the accession of Kashmir to Pakistan he sacked him. In his place he appointed Justice Mehr Chand Mahajan as the Prime Minister on 15 October 1946. The drama of accession of Kashmir to India comes out vividly in his memoirs 'Looking Back', Asia Publishing House, New Delhi, 1963. His account is corroborated by the books listed in the bibliography. **It is best to summarise the major events :**

- To gain time, the Maharaja signed a standstill agreement with Pakistan in August and was negotiating a similar arrangement with India.
- Towards the end of September, Pakistan started a blockade of the State by preventing vehicles carrying food, clothing, petrol, ammunition and other daily necessities from entering the State via Sialkot or Kohala. It was designed to coerce the State to accede to Pakistan and was a violation of the standstill agreement signed by Pakistan and Jammu and Kashmir.
- As per Liaquat - Akbar plan, revolt erupted in Poonch area after 4,000 rifles were delivered there, raids from

West Punjab into Western Jammu region also gathered momentum.

- Preparation of Tribal Invasion continued under the supervision of Khan Abdul Qayyum Khan, Chief Minister of NWFP. Jinnah was informed of the planned invasion a few days before the D-day on 20 October and he was requested to be in Abbottabad for a triumphal drive into Srinagar. It appears that Jinnah did not know of this plan in advance when he had sent his Kashmiri private secretary, Mr. K.H. Khurshid, ostensibly to visit his family in Srinagar. In fact "he had been busy in the State for the last three months creating a strong communal agitation to force the Maharaja to accede to Pakistan."[15] Khurshid (who was later appointed as the President of Pakistan-occupied Kashmir and held this post for four years) was trapped during the invasion, jailed and repatriated only after ceasefire on January 1, 1949. He records the following about the invasion :

"On my return from Kashmir in 1949, when I met Liaquat Ali in January of the same year in Lahore, I asked him whether the Qaid-e-Azam was informed about this plan. He did not answer but responded with a weak nod of assent."[16] Khurshid's view is corroborated by the Governor of NWFP George Cunnigham. On October 26, 1947 he wrote in his diary "Apparently Jinnah himself first heard of what was going on about 15 days ago (10 Oct ?), but said "Dont tell me anything about it. My conscience must be clear."[17]

It is widely believed that the failure of the Tribal Invasion to capture Srinagar deeply shocked Mr.Jinnah resulting in rapid decline in his health - "He never recovered from this shock," recalls Khurshid.

The view that "Tribal invasion was a criminal folly which sealed the fate of Kashmir"[18] means that the dithering Maharaja was forced off the fence of indecision to accede to India. Mr Jinnah was a man of great integrity. This folly was a major event which must have contributed to his dying declaration that :

"Pakistan was the greatest blunder of my life"[19].

- 29 September 1947, Sheikh Abdullah was released after being in Jail since 20 May 1946, essentially to help in maintaining communal harmony in the State after the revolt broke out in Poonch as per the Liaquat-Akbar plan.
- 15 October, Mehr Chand Mahajan took over as the Prime Minister. He replaced Mr.Ram Chandra Kak, "who made no secret of his pro-Pakistan leanings. He had been, it is said, hobnobbing with Pakistani politicians promising them Kashmir on a platter."[20] Soon after Mahajan took over Major Shah who was a Secretary in Pakistan government and the personal envoy of Jinnah, called on him. He had come to Srinagar in the first week of October to exploit the blockade and negotiate terms of accession. He had met R.L. Batra the Deputy Prime Minister and General Janak Singh with no results. Shah was the son-in-law of Sir Abdul Rashid, Chief Justice of the Lahore High Court and was well known to Mahajan who was a member of the bench of the same court. Shah urged at length to extract a promise from Mahajan to advise the Maharaja to accede to Pakistan. Mahajan agreed to consider the matter with an open mind on merits, provided the blockade was lifted. Shah felt it was reasonable. He went away to consult Jinnah telegraphically and met Mahajan again that afternoon. He conveyed Jinnah's message that lifting of blockade

was not possible till the matter of accession was decided and that Mahajan should go to Lahore to discuss the question of accession. Mahajan declined this coercive proposition. Shah became a bit aggressive and told Mahajan that refusal to accede immediately might result in serious consequences. The Tribal Lashkars were, on 15 Oct, moving into their launching pads. To this Mahajan replied that "a matter of this kind would throw the State into the lap of India."

- Night 20/21 October, the Tribal Invasion launched.

- On 24 October, the Deputy Prime Minister Mr.Batra left Srinagar for Delhi carrying a letter of accession to India from the Maharaja and a personal letter to Pandit Jawaharlal Nehru and another to Sardar Patel asking for military help. Mehr Chand Mahajan also wrote to both requesting them to save the State from Pakistan's unprovoked aggression.

- On 25 October 1947, Sheikh Abdullah left for Delhi and sent Bakshi Ghulam Mohammed and G.M Sadiq to Pakistan to meet Jinnah and Liaquat Ali, but they were not received.

- By 25 October the rape of Baramula started. The orgy delayed the invaders in reaching Srinagar.

- 25 October, V.P. Menon was sent to Srinagar to assess the situation. Saw the "stillness of graveyard"[21] in the city.

- The Maharaja was persuaded by V.P. Menon and Mahajan to leave Srinagar for Jammu at 0200 hrs. on 26 October, on the plea that it may be necessary to negotiate accession terms and it would be more convenient if he was in Jammu.

- On 26 October, V.P. Menon and Mehr Chand Mahajan flew to Delhi and drove straight to the Prime Minister's house on the York Road (Now Motilal Nehru Marg). Mahajan records :

"The Prime Minister and Sardar Patel were there and were apprised of the situation that had arisen. In view of the advance of the raiders towards the town of Baramula and Srinagar, I requested immediate military aid on any terms. I said somewhat emphatically that the town of Srinagar must be saved at any cost from loot and destruction. During the talk, the Prime Minister said that even if the town was taken by the tribesmen, India was strong enough to re-take it. Its recapture, however, could not have undone the damage that would have resulted. I, therefore, firmly but respectfully insisted on the acceptance of my request for immediate military aid. The Prime Minister observed that it was not easy at the spur of the moment to send troops as such an operation required considerable preparation and arrangement, and troops could not be moved without due deliberation merely on my demand. I was, however, adamant in my submission; the Prime Minister also was sticking to his own view. As a last resort, I said, "Give us the military force we need. Take the accession and give whatever power you desire to the popular party. The army must fly to save Srinagar this evening or else I will go to Lahore and negotiate terms with Jinnah." When I told the Prime Minister of India that I had orders to go to Pakistan in case immediate military aid was not given, he naturally became upset and in an angry tone said, "Mahajan, go away." I got up and was about to leave the room when Sardar Patel detained me by saying in my ear, "Of course, Mahajan, you are not going to Pakistan." Just then, a piece of paper was passed over to the Prime Minister. He read it and in a loud voice said, "Sheikh Sahib also says the same thing." It appeared that Sheikh Abdullah had been listening to all this talk while sitting in one of the

bedrooms adjoining the drawing room where we were. He now strengthened my hands by telling the Prime Minister that military help must be sent immediately. This came as a timely help for the success of my mission to New Delhi. The Prime Minister's attitude changed on reading this slip. After a few minutes talk, he told me to go and have some rest at Sardar Baldev Singh's house. He was calling a meeting of the Defence Council at 10 A.M. to discuss the matter and promised to convey its decision to me through my host, Sardar Baldev Singh (the Defence Minister), before lunch."[22]

- 27 October, Jinnah flew from Karachi to Lahore. Met Liaquat who was confined to bed. Appreciating that Liaquat Ali had messed up the whole affair, he asked the Punjab Governor Sir Francis Mudie to order General Douglas Gracy, the British C-in-C of Pakistan, to use Pakistan Army to secure Jammu and the Srinagar airfield. Gracy demurred. In any case, it was too late because Indian troops were landing at Srinagar since the morning of 27 October 1947.

- On 1 November 1947 Mountbatten flew to Lahore to meet Jinnah to resolve the conflict in Kashmir. He presented the formula. 'The Governments of India and Pakistan agree that where the ruler of a state does not belong to the community to which the majority of his subject belong and where the State has not acceded to that Dominion whose majority community is the same as the State's, the question whether the state should finally accede to one or the other of the Dominions should in all cases be decided by an impartial reference to the will of the people'. Jinnah replied that a plebicite was redundant and undesirable. He also refused to include Hyderabad in the reckoning.[23] Jinnah then proposed to Mountbatten, there should be a simultaneous withdrawal of Indian troops and Tribal invaders. On beingasked how anyone could guarantee that the latter would also be withdrawn, Jinnah replied "if you do this, I will call the whole thing off"[24]. The meeting failed and Kashmir problem took firm roots.

It was Liaquat Ali who stated that the accession of Kashmir to India was the result of a conspiracy and fraud. Quite obviously, it was an alibi for the failure of his plan to capture the State.

What emerges from this fateful drama is that the accession of the State of Jammu and Kashmir was not based on any formula, principle, plan or logic. The Indian Independence Act gave absolute right to the rulers of the princely states. In fact, it was Jinnah who forcefully insisted and supported the absolute right of the rulers to exercise their choice. In this stand he had an eye on the Nizam of Hyderabad and Nawab of Bhopal. Nor was accession based on any conspiracy or fraud. It was simply the outcome of the *'criminal folly'* of launching the Tribal Invasion. In human affairs it is easy to be swept away by emotions and resort to violence, but it is impossible to see the end. That is why the recurring lesson of history is that violence solves no problems. Despite this, man in his emotional myopia persists in the use of force. That was precisely what happened in Jammu and Kashmir.

Many people fault Nehru for making an offer to finally settle the future of the State by reference to the people. Under the Indian Independence Act, it was not necessary to do so. Others fault Mountbatten for having insisted on this provision. The truth is that this offer was made in the hope that it would also help solve the problem of Junagarh and Hyderabad States. The assertion of Mountbatten about Kashmir on this score is crystal clear :

"Consistent with their policy that, in the case of any State where the issue of accession has been the subject of dispute, the question of accession should be decided in accordance with the wishes of the people of the State, it is my Government's wish that as soon as law and order have been restored in Kashmir and its soil cleared of the invader the question of the State's accession should be settled by reference to the people"[25].

The *truth* is that the invader[26], after 50 years is still in Jammu and Kashmir; time has marched on including two more wars fought between India and Pakistan culminating in the Simla Agreement signed by the two sides in 1972 as an important landmark in the history of South Asia.

* Despite all the care for secrecy, information about the invasion leaked out to India - particularly when plans to muster transport to carry the Tribal Lashkars to various points on the border of Kashmir were formulated. Shiv Saran Das was the Deputy Commissioner at Dera Ismail Khan, the Southern district of North West Frontier Province. He came to know of the plan. He came on leave to India in late September and informed the Government, it is on the files. Similarly, Major Onkar Singh Kalkat who was a Brigade Major at Bannu also came to know of it and gave the information to the Indian Army. However, nothing could be done as J&K was an Independent State.

NOTES

1. Sisir Gupta in *Kashmir : A Study in India Pakistan Relations,* Asia Publishing House, New Delhi - 1966.
2. Ibid
3. Dr. Rafiq Zakaria in *Iqbal : The Poet and Politician.*
4. Ibid
5. Ayesha Jalal in *Sole Spokesmen.*
6. K.H. Khurshid in *Memoirs of Jinnah,* Oxford University Press, Karachi, 1990.
7. Trench Charles Cheevix in *Frontier Scouts,* Butter and Tanner Ltd., London, 1985.
8. Datta Khel is important centre of Tochi Scouts, South West of Peshawar.
9. Trench
10. Ibid
11. Ibid
12. Sir Olaf Caro in *Wells of Power,* Macmillan and Company, 1951.
13. Mehr Chand Mahajan, Chief Justice of India, in *Looking Back and Autobiography,* Asia Publishing House, New Delhi, 1963.
14. Professor C.H. Philip, *Partition of India : Policies and Perspectives 1935-47.*

15. Mehr Chand Mahajan

16. K.H. Khurshid

17. Quoted by A.G. Noorani in his review of the book *Kashmir : Disputed Legacy* by Alastair Lamb in Frontline Magazine (of Hindu newspaper) of 27 March 1992.

18. On 25 November 1991, there was forty five minutes discussion held in the external service of the BBC, on Kashmir. The participants were:

    (a) Mr Mark Tully the BBC representative for the South Asia for over 20 years.

    (b) Professor Alstair Lamb, the author for the book, 'Kashmir: Disputed Legacy'.

    (c) Dr Wiqar Ahmed who taught history in Karachi University before he migrated to the U.K and joined the Urdu service of the BBC.

    During the discussion, Dr Wiqar Ahmed quoted from the book "Memories of Jinnah" by his secretary K.H. Khurshid that the "Tribal invasion was a criminal folly which sealed the fate of Kashmir". The point he was making was that the Maharaja was forced by Pakistan to accede to India to survive as also to save Srinagar from loot and rape as meted out to Baramullah.

19. In its issue of 23 December 1996, The Time Magazine devotes a full page to the founder of Pakistan, Mohmmad Ali Jinnah. Quoting his doctor, the magazine records Jinnah's final judgement on his achievement of creating Pakistan in the following words:

    "Pakistan" he said, has been "the biggest blunder of my life".

Jinnah was a man of great moral courage and integrity. His sense of justice and his dream of Pakistan as a modern nation must have been shattered by what happened after partition.

The following three factors must have weighed heavily in his tragic self-judgement.

* He articulated his dream about Pakistan in his address to the constituent assembly of Pakistan in Karachi on 11 August 1947: "You are free to go to your temples, you are free to go to your mosques or to any other place of worship in this State of Pakistan. . . You may belong to any religion or caste or creed - - that has nothing to do with the business of the State. . . . We are starting in the days when there is no discrimination, no distinction between one community and another, no discrimination between one caste or creed and another. We are starting on this fundamental principle that we are all citizens and equal citizens of one state. The people of England in course of time had to face the realities of the situation and had to discharge the responsibilities and burdens placed up on them by the government. . . . Today, you might say with justice that Roman Catholics and Protestants do not exist; what exists now is that every man is a citizen, an equal citizen of Great Britain. . . All members of the Nation. . . . We are starting with this fundamental principle that we are all citizens and equal citizens of one state. Now I think we should keep that in front of us as our ideal and you will find that in course of time Hindus would cease to be Hindus and Muslims would cease to be Muslims, not in the religious sense, because that is the personal faith of each individual, but in the political sense as citizen of State. . . I shall always be guided by the principles of justice and fairplay without any, as is put

in the political language, prejudice or ill-will, in other words, partiality or favouritism. My guiding principle will be justice and complete impartiality, and I am sure that with your support and cooperation, I can look forward to Pakistan becoming one of the greatest Nations of the World. . . ."

* When Muslims, Hindus and Sikhs turned into animals and started butchering one another in India as well as Pakistan immediately after 14/15 August 1947, triggering the biggest migration of people in History, the sensitive lawyer must have been deeply hurt.
* The final blow to his faith in humanity must have been by the Tribal invasion of Kashmir in which Liaqat Ali and his lieutenants who organised the invasion aranged with the tribesmen that:

  "their only remuneration was to loot non-Muslims."

20. Mehr Chand Mahajan
21. V.P. Menon in *The Story of Integration of the Indian States,* Macmillan, New York, 1951.
22. Mehr Chand Mahajan
23. A.G. Noorani quotes this from *'Sardar Patel's Correspondence 1945-50'* in his review of the book 'Kashmir: Disputed Legacy' by Professor Alstair Lamb in Front Light Magazine of 27 Match 1992.
24. Campbell Allen Johnson in *Mission with Mountbatten,* Roberthale, London, 1951.
25. Sisir Gupta.
26. Lt. Gen. Sir James Wilson and Lt. Gen M.L. Chibber, in 'Jammu and Kashmir Problem: The Truth', Parts I and II in April – June 97 and July – September 97 issues of the Journal of the United Services Institute of India. The following extract from Chapter seven of Wilson's

(who was military assistant to General Douglas Gracy, the C-in-C of Pakistan during 1948) memoirs throws light on this point:

"It remained, however, vital for Pakistan somehow to defend the mainly Muslim areas of Kashmir ourside the vale itself. A fiction, therefore, arose by which this territory became the responsibility of a local militia, known as the Azad (Free) Kashmir Forces, commanded by a regular Pakistani officer, who used the 'nom de guerre' of General Tariq. Elements of the Pakistan Army found their clandestine way into the Azad Kashmir Forces in sufficient numbers to stablise the military situation. British officers were strictly prohibited from taking part in these operations, and were not permitted to enter Kashmir. Both Messervy and Gracy were aware of the progressive reinforcement of the Azad Kashmir forces from the regular Pakistan Army. Neither, however, gave any instructions or advice to General Tariq, who dealt directly with Iskander Mirza, the Defence Secretary, and through him with Liaquat Ali Khan, the Prime Minister who also held the Defence portfolio.

Though this covert involvement of the Pakistan Army in Kashmir, byzantine in its complexity, was clearly not satisfactory, in practice it operated surprisingly well".

CHAPTER 3

Ladakh : Geography, History and Culture

GEOGRAPHY

Beyond the Great Himalayas lie the mountains and high-altitude plateaus of Tibet. Millions of years ago, this region was the bed of an ocean, known to geologists as the Sea of Tethys. It was pushed up by the massive force of tectonic movement as the southern continent of Gondwanaland, which detached from the African landmass drifted northwards and rammed into the Eurasian landmass. The Great Himalayas, representing the earth and rock that was pushed up (and is still rising) along the line of contact, forms for three quarters of its length - a 2,400 km (1,500 mile) sweep from southeast to northwest - the southern rampart of the Tibetan plateau. To the south, ranges of lesser altitude run parallel to it, and northwards the valleys and plateau stretch away for hundreds of miles. At its western end, however, the Great Himalayas branch out into numerous ranges and subranges. The Dhauladhar and the Pir-Panjal, no doubt, are a continuation of the foothill chains on a grander scale. But, to the north it acquires a major subsidiary, the Zanskar Range, some 480 km (300 miles) long, and nearly 160 km (100 miles) across, running from the Shipki-la to the Suru Valley. Parallel to this range to the North are two more ranges, belonging to a different geological system : the Ladakh Range and the Karakoram, second in height and grandeur to the Great Himalayas itself. The major ranges, routes as they were in 1947, and places are shown on sketch.

Much of Ladakh, being in the rain-shadow of the Great Himalayas, is high-altitude desert. Cultivation is possible only where glacier-fed streams can be tapped for irrigation. The villages are scattered and communication possible only via mountain passes varying in height between 3,350 metres (11,000 feet) and 5,500 metres (18,500 feet). La meaning a pass ; Ladakh stands for the land of passes. In winter, when river waters freeze, the gorges through which they flow constitute an easier pathway for the traveller.

The Zojila, the most important of the passes between Kashmir and Ladakh, gives little sign that it is cutting through the world's mightiest range. Here, on Ladakh's southern boundary, the peaks of the Great Himalayas are modest compared to the giants of its central portion, or with its western-most sentinel, the Nanga Parbat. The exception is the massif of Nun-Kun, whose twin peaks (7,135 metres or 23,409 feet, and 7,087 metres or 23,250 feet) tower high over their neighbours. The famous shrine of Lord Shiva at Amarnath is only a few miles to the east of the Zojila (original name Seojila, named after Shiva).

The Great Himalayas is separated from the Zanskar Range by the Suru and Zanskar valleys. Together these form a trough ; its lowest point 2,600 metres (8,500 feet) at the confluence of the Suru river with the Indus; (The Suru joins Shingo, and the Shingo joins Indus).

In 1947, Kargil on the banks of Suru River, was an important meeting point and trade centre, being equidistant at about 200 kilometres (125 miles) from Srinagar in Kashmir, Padam (the central village of Zanskar), Leh in Ladakh, and Skardu, the principal town in Baltistan which used to be the winter headquarters of the Ladakh tehsil.

The Zanskar Range is an incredibly convoluted complex of mountains, with peaks varying in height between 5,500 metres (18,000 feet) and 6,400 metres (21,000 feet) and gashed in all directions by the valleys of streams which drain into the Zanskar and the Indus rivers. The logic of its configuration is not easily apparent on the ground ; all the more remarkable, then, the achievement of the traders who worked out the routes along the valleys and over the passes centuries ago. The autumn trade-route between Zanskar and Leh is said to have involved crossing the Khara river, a tributary of the Zanskar, no fewer than 108 times. [The winter passage to Leh is easier and shorter because the Zanskar river freezes, and is known as 'Chaddar' i.e. sheet (of ice).]

The valley of the Indus river, runs through Ladakh. Its source is near the massif of holy Mount Kailash in Tibet, it enters India at Demchok and has no major tributary until it takes the waters of the Zanskar. The Indus valley is historically Ladakh's heartland and stretches from Upshi, where the trade route through the Zanskar Range and across Rupshu from Manali comes in, down to Khalsi where it is joined by the trail from Kashmir.

Ladakh was an independent Kingdom from the early 10th century until 1834, and at various times the capital was at Shey, Basgo and Timisgam. It was Sengge Namgyal, greatest of the Ladakhi Kings, who settled in Leh in the early 17th century. Apart from those in Zanskar and Nubra, all of Ladakh's major *gompas* are in or near the Indus valley. They number 13 in all, and several of them, like Hemis and Chemrey, were founded under royal patronge. Some others, according to the Ladakh Chronicle, were not built by human hands at all, but 'came into existence by a miracle'.

To the north of Indus valley rises the Ladakh Range, an extension of Kailash Range which has its origin in the massif of Mount Kailash, far away in Western Tibet. In Ladakh, it makes the divide between the rivers Indus and Shyok, ending at their confluence in Baltistan. With a fairly uniform elevation of 5,600 - 6,000 metres (18,500 to 19,500 feet), none of its peaks is more than 6,250 metres (20,500 feet). Most passes of the Ladakh Range like the Changla (5,357 metres or 17,574 feet) and the Chorbatala (5,089 metres or 16,696 feet) are little more than dips, sometimes descending a mere 200-300 metres below the general level of the crest. Yet all, particularly the glaciated Khardung-la, and Digar-la were regularly crossed by the traders and their animals for centuries.

In winter, the Yarkand-bound caravans preferred to take the Chang-la, and go up the Shyok gorge which was possible only when the river was frozen and became negotiable. But, most of the traffic was in late summer, and went via the Nubra valley route over the Digar-la or the Khardung-la and the Saser-la. This was the last inhabited area before the Karakoram crossing, and it would be ten or twelve days before cultivation was reached on the other side. Panamik, the last village, was a regular halting place, where men and animals might be recruited for the crossing, or recuperate after it. An interesting practice was the use of sheep carrying about 20 kg of grain, who accompanied the caravans. The grain and the sheep provided food during the passage without diminishing the commercial cargo. All the fields were protected from the animals of the passing caravans by thick thorn hedges; and the local people made a profitable income from the supply of grain and fodder. Even the mighty Karakoram, Ladakh's northernmost rampart, did not remain inviolate to commerce. But, it extracted a terrible price. For several days' march on either side of the Karakoram pass (5,575 metres or 18,290 feet), the route was marked by the bleached bones of humans and animals (Yaks, ponies, bactrian camels and sheep) that had succumbed to altitude and exhaustion. It is said that no merchant or pony driver set off from Yarkand (in Sinkiang, China) without including a shroud among his personal effects.

The establishment of a Tibetan Kingdom in the tenth century substantially altered the demographic nature of the earlier population of the Mons (meaning Southerners), who were the dominant group of people of Indian ethnic origin. Later arrivals from Kashmir and Baltistan settled mostly in the Dras and Suru valleys and Purig, creating a people of mixed origin - Indian, Dardic, Mongol and Balti.

As the centuries rolled on, another religious force gradually began to make itself felt. From its bastion in Kashmir, where it had all but eclipsed the Hinduism that had superseded Buddhism, Islam began to make inroads into northwest Ladakh. Many of those who preached it seem to have been Shias, for it is the Shia sect that is predominant in the Kargil area. This was a peaceful penetration; but, various attempts were made to impose Islam by the sword. Central Ladakh was subject to invasions from Kashmir, Central Asia and Baltistan. In this last incident, about 1600 AD, the King Jamyang Namgyal, was defeated and taken prisoner to Skardu. Shortly after, he was released and restored to his throne on condition that he would marry the Muslim Skardu King's daughter later known as Khatun Gyalmo. This curious episode has been invested in folk memory with a romantic gloss as an affair of the heart between Jamyang and the Balti princess, followed by a secret marriage, or perhaps an illicit pregnancy. Either way, the Ladakhis turned the situation deliciously to their advantage by recognising in their Muslim queen an incarnation of the Goddess Tara. The son of the union, Sengge (modification of Singh i.e. lion) Namgyal, turned out to be the most famous king of Ladakh, who extended the Kingdom's boundaries to their farthest limits. More religious structures were built under his patronage than in any other reign. These included the monasteries of Hemis, Chemrey and Hanle (far away near the border with Tibet), a number of *mane* (walls faced with prayer stones) near Leh and in Zanskar, and a gigantic gilded Buddha image at Shey. Sengge was also the builder of the nine-storey palace in Leh.

During the reign of his successor, Deldan Namgyal, Sengge's conquests to the east were lost as a result of a disastrous war with Tibet. When the Tibetan army invaded and besieged the Ladakhis at Basgo, it was only by appealing to the Mughal Governor of Kashmir for help that Deldan was able to drive them out. The price he paid for this was acceptance of Mughal suzerainty. The Treaty of Timisgam, a tripartite agreement by which all outstanding issues were settled at the end of the war, had as its most pregnant clause the stipulation that all the Pashm produced in western Tibet should be reserved for Kashmir and laid down the rules and procedures for the trade, via Ladakhi and Kashmiri middlemen.

It was indeed the lucrative trade in this luxury article which, exactly a century and a half later, tempted the ruler of Jammu, Gulab Singh, to send his most valiant general, Zorawar Singh, to invade and conquer Ladakh in 1834, and eight years later to make an attempt on Tibet itself in

Ri zong Monastery near Leh twice plundered by Pakistanis

a bid to lay hold on the trade from its very source. This latter attempt was a failure and Zorawar was defeated and killed. But, shortly afterwards Gulab, as a result of the agreement with the British which made him Maharaja of Kashmir, became heir to the Kashmir monopoly of the material. Thus, it was, that the trade in Pashm not only shaped the events of the period, but was also responsible for the political configuration of the State of Jammu and Kashmir as we know it today.

HISTORY

For close on 900 years, from the middle of the tenth century, Ladakh was an independent Kingdom, its dynasties descending from the kings of old Tibet. Its political fortunes ebbed and flowed over the centuries, and the kingdom, was at its greatest in the early seventeenth century under the famous king Sengge Namgyal, whose rule extended across Spiti and western Tibet upto the Mayumla beyond the sacred sites of Mount Kailash and Lake Mansarovar.

And gradually, perhaps partly due to the fact that it was politically stable, in contrast to the lawless tribes further west, Ladakh became recognized as the best trade route between the Punjab and Central Asia. For centuries it was traversed by caravans carrying textiles and spices, raw silk and carpets, dyestuffs and narcotics. Heedless of the land's rugged terrain and apparent remoteness, merchants entrusted their goods to relays of pony transporters who took about two months to carry them from Amritsar to the Sinkiang towns of Yarkand and Khotan. On this long route, Leh was the half-way house, and developed into a bustling entrepot, its bazars thronged with merchants from far away countries.

The famous Pashm (better known in the west as Cashmere) also came down from the high altitude plateau of eastern Ladakh and western Tibet where it was produced, through Leh to Srinagar, where skilled artisans transformed it from a matted oily mass of goat's underfleece into shawls known the world over for their softness and warmth. Ironically, it was this lucrative trade that finally spelt the doom of the independent kingdom. As stated earlier it attracted the covetous gaze of Gulab Singh, the ruler of Jammu in the early 19th century. There followed a decade of war and turmoil, which ended with the emergence of the British as the paramount power in North India. Ladakh, together with the neighbouring region of Baltistan, was incorporated into the newly created State of Jammu & Kashmir. Just over a century later, this union was disturbed by the partition of India.

TRADE ROUTES

For all its seeming inaccessibility, Ladakh's position at the centre of a network of trade routes traditionally kept in constant touch with the outside world. From Chinese Central Asia, the mighty Karakoram range was breached at the Karakoram pass, a giddy 18,350 feet (5,600 m). The trail from Yarkand crossed five other passes, of which the most feared was the glacier-encumbered Saser-la north of Nubra. Travellers from Tibet could take one of two main routes. From the central part of the country, the Tsangpo Valley, they could pass the holy sites of Kailash-Mansarovar and reach Gartok, on a tributary of the upper Indus, from where they followed the river down to Leh. Trade with the Pashm producing areas of Western Tibet followed by a more northerly route, taking in the village of Rudok, a few miles into Tibet, and from there across to Chushul near the Pangong-tso, up the length of the lake to Tangtse, then across the 18,300 feet (5,578 m) Changla to the Indus, and on to Leh. Baltistan, joined administratively with Ladakh for 100 years, was linked to it either via the Indus up to its confluence with the Suru-Shingo river, and on upto the Kargil; or by the Chorbat-la pass over the Ladakh range, the trail dropping down to the Indus 40 km below Khalsi, and following the river up to Leh.

The two main approaches to Ladakh from south of the Himalayas are roughly the same as today's motor roads from Srinagar and Manali. The merchants and piligrims who made up the majority of travellers in the pre-modern era, travelled on foot or horseback, taking about 16 days to reach Srinagar, though a man in a hurry, riding non-stop and with changes of horse arranged ahead of time all along the route, could do it in as little as three days. The mails, carried in relays by runners stationed every four miles or so, took four or five days.

CENTRAL LADAKH - the Cultural Heartland

The geographical backbone of Ladakh, the Indus valley, particularly from Upshi down to Khalsi, is also the region's historic heartland. All the major sites connected with the former kingdom's dynastic history are here, starting with Leh, the capital city since the early seventeenth century when Sengge Namgyal built his nine-storey palace. A few kilometres up the Indus is Shey, the most ancient capital, with its palace and temples, their vibrantly coloured murals cleaned and restored in the mid 1980s. Down the river, Basgo, right on the road and Timisgam, a short way up a side valley as capital cities, when the country was temporarily divided into two parts in the fifteenth century, and both have the remains of forts and temples dating from the period of their brief glory. Stok, just across the river from Leh, is the village with which the deposed royal family was compensated for the loss of its throne after Zorawar Singh's invasion. Its palace houses a museum of artefacts associated with the dynasty and there is also a small *gompa*.

Partly as a result of royal patronage, the central area of Ladakh has the greatest concentration of major *gompas*. Of the twelve situated on or near the Indus, the oldest is Lamayuru also known as Yung Drung (Swastika), believed to have been a sacred site for the pre-Buddhist religion known as Bon. Phyang, Hemis and Chemrey were all founded under the direct patronage of members of the ruling Namgyal dynastry. Phyang represents an act of penance by the sixteenth century king Tashi Namgyal for the violence and treachery through which he reached the throne. It enjoyed the status of sanctuary for fugitive criminals, by royal decree. Hemis, together with Hanle near the Tibetan border, was founded at the instance of king Sengge Namgyal, and Chemrey by his widow as a posthumous act of merit for him. Stakna, dating from a slightly earlier period, was endowed by the Namgyal kings at various times. All these belong to the Red Hat Kar-gyuk-pa sect of Tibetan monasticism.

The reformist Ge-lugs-pa (Yellow Hat) sect is also well represented in central Ladakh by Thikse, Likir, Rizong and Spituk, the last of which has daughter houses* at Stok, Sabu and Sankar. Rizong, the only *gompa* which is not as yet approachable by a motorable road, is situated a few kilometres up a side-valley at Uley-Tokpo. It was founded only a century and a quarter ago by a devout layman-turned-Lama, with the purpose of giving full expression to the strict monastic rule of the Ge-lugs-pa. While the paintings and images in its temples may, to some extent, lack the aesthetic and antiquarian interest of those in the older establishments, this *gompa* nevertheless, has an indefinable atmosphere of peace and dedication which reflects faithfully the inwardness of the Buddhist way.

The smaller but much older Nying-ma-pa and Saskya-pa monastic sects are represented respectively by Tak-Thok and Matho *Gompas*. Tak Thok, at the foot of the Chang-la, incorporates one of the many caves in the Himalayas, where the Indian Buddhist apostle Padamasambhava is said to have rested and meditated on his journey to Tibet.

*Subsidiary gompas

Matho *Gompa* a slightly run-down structure, but a vibrant religious community. It is famous for its festival of the oracles which takes place early in the year, usually in late February or early March.

But the jewel among central Ladakh's religious sites is Alchi. Abandoned centuries ago, as a place of regular worship, it has been lovingly maintained by the monks of Likir, the nearest functioning *gompa*. Known as Chos-Kor, or religious enclave, it comprises five temples, the richest in paintings and images being the Dukhang (assembly hall) and the three-storey Sum Tsek. Its murals, dating from the eleventh and twelfth centuries, predate the Tibetan style of painting that is present in all the other *gompas*. Some of them are reminiscent of the paintings of the far-off Ajanta Caves and presumed to be almost the sole survivors (along with some in Phugtal Gompa in Zanskar, and Tabo in Spiti) of the Buddhist style prevalent in Kashmir during the first millennium AD.

LEH: The Capital City

Leh, the main town of the region, is dominated by Sengge Namgyal's nine-storey palace, a building in the grand tradition of Tibetan architecture, said to have inspired the famous Potala in Lhasa, which was built half a century later. Above it, on Namgyal Tsemo, the peak overlooking the town, are the ruins of the earliest royal residence at Leh, a fort built by King Tashi Namgyal in the sixteenth century. The associated temples remain intact, but they are kept locked except during the morning and evening hours, when a monk toils up the hills from Sankar Gompa to attend to the butter lamps in front of the images. At the foot of the edifice is a mosque, originally built to honour Sengge's Muslim mother, Khatun Gyalmo.

Down in the bazar, the main sites are the Jo-Khang or Gompa Soma, a modern ecumencial Buddhist temple, and the imposing mosque dating from the late seventeenth century almost opposite. But, the pleasure of Leh is not confined to the purposeful visting of sites. A particularly charming sight is the line of women from nearby villages sitting along the edge of the footpath with baskets of fresh vegetables brought for sale to the town's people. Chhang Gali, behind the main bazar, is less bustling, but has intriguing little shops selling apart from Chhang, the local brew, curios and jewellery, and further on, are the labyrinthine alleyways and piled up houses of the old city, clustering around the foot of the palace hill. In the other direction, down from the bazar, are the stalls of the Tibetan traders where you can bargain for pearls, turquoise, coral, malachite lapis lazuli and many other kinds of semi-precious stones and jewellery, as well as curiously carved Yak-horn boxes, quaint brass locks, china or metal bowls, or any of a whole array of curios.

MAJOR MONASTERIES

Shergol: The Shergol gompa is about 237 km from Srinagar and is situated between Kargil and Mulbekh, at the very mouth of the valley of the Wakha river. Shergol provides a spectacular view of the Wakha valley from the top of the gompa. It is truly a gompa in the literal sense (Gompa : a solitary retreat) as nowhere else can one experience such peace and quiet as at Shergol.

Mulbekh: Mulbekh has two main gompas – Serdung, which belongs to the drug-pa sect and is looked after by the monks of Chemre, and Gandestse, which comes under the Ge-lugs-pa monks of Likir. Built high on a rocky hill-top, the latter involves a laborious walk. Mulbekh is a very important centre on the Srinagar-Leh road as it marks the dividing line between the Muslim belt of Kargil and Buddhist central Ladakh. From the road itself one can see the seven-metre high statue of Maitreya or Chamba (the Buddha to come in a standing posture hewn out of a single boulder).

Lamayaru: The Lamayaru monastery is perched atop a rocky spur that suddenly juts out from the floor of the valley. Situated a little off the Srinagar-Leh route (on the right) as it descends after crossing the Fatu-la, this site is presumably the oldest in the area. It presents a spectacular view of strange undulations of lacustrine clay from a drained lake bottom, hence the name 'moonland' given to it. The old monastery is in ruins, another built much later, is the one in use. However, the old building does have some frescoes of Mandalas and fearsome divinities with their consorts and images of Vairocana and other Buddhas and has much higher antiquarian significance.

Belonging to the order of the Dri-gung-Kargyugpa, a sub-sect of the 'Red Hats', Lamayaru was one of the few monasteries where criminals could traditionally seek asylum, thus its other name, Tharpa Ling (place of freedom). The *dukhang* (assembly hall) built some time in the sixteenth century, is situated next to a cave where the Tibetan sage Marpa had meditated. With torchlight one can see the life-size statues of Marpa and his chief disciple, Milaraspa, inside the cave.

Rizong: About 32 km from Khaltse, the Ri-dzong monastery is situated well off the beaten track. A side track leaves the main route and leaves the river Indus and through the village of Uley-Topko, going upstream by the side of a tributary till the *chamoling* (nunnery). From here on there is only a footpath, and it can easily take an hour to reach the gompa.

Rizong is a recently built retreat (a hundred and five years old) belonging to the Ge-lugs-pa order and is probably the only place where monastic discipline is adhered to in its orthodox form. About 30 Lamas live here and they follow a life of strict discipline. The Chomos (Buddhist nuns) are not allowed in the main monastery except for special prayers. Most of their life is spent within the four walls of the chomoling, performing daily chores and in prayers.

Alchi: This is one of the oldest monasteries in Ladakh, dating back to the eleventh century AD, when the great translator from western Tibet, Rinchen Zangpo, returned from India having acquired extensive tutelage in Buddhism. Thus Alchi has a distinct Indian style of architecture and decor, and is in fact the only remnant of Indian Buddhism in Ladakh.

The Alchi Monastery is situated a little off the Srinagar-Leh route, across the Indus, about 70 km short of Leh and a couple of kilometres short of Saspol. Unlike many monasteries, Alchi is built on a flat terrain and not atop a cliff or hill. Architecturally, it seems disarranged, consisting of a number of courtyards connected by narrow corridors, as in a maze. The Sum-Tsek is a three-storey temple. Every inch of space on its walls is covered by murals full of vibrant colours in remarkable detail. There is a passage in the ceiling of the ground floor to accommodate the enormous images of the three Boddhisatvas : Avalokiteswara, Manjusri and Maitreya. Through a courtyard one can approach the *dukhang* or assembly hall, which has a statue of Avalokiteswara, believed to be made of pure gold. The central figure is that of Vairocana situated on a high pedestal behind the sacrificial altar. There are figures of *apsaras* (female angels) and the walls have paintings of various mandalas.

This five templed gompa is not an active monastery any more. Though it was once a 'Red Hat' monastery, today the Ge-lugs-pa (Yellow Hat) monks of the Likir monastery look after it.

Likir : Likir lies five kilometres off the main road between Saspol and Basgo and is built over a low flat hill.

SIEGE OF SKARDU
11 FEB - 14 AUG 1948

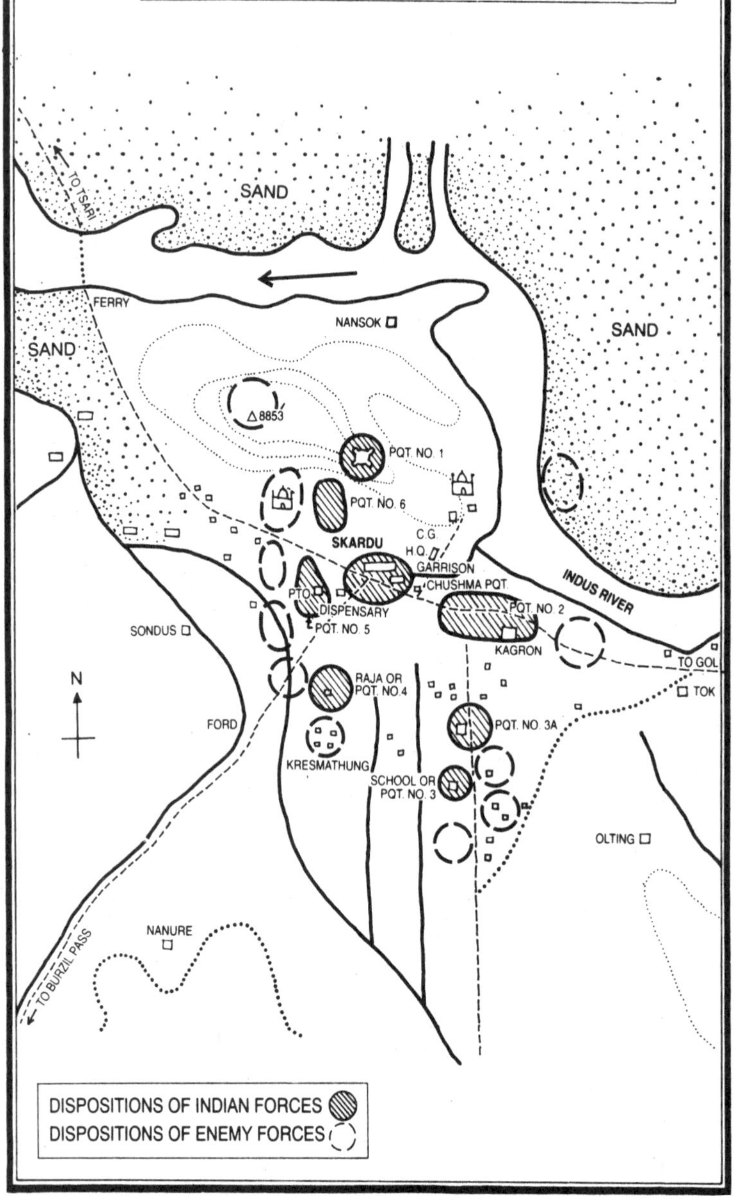

SKETCH - 3

Though it was originally built by the followers of the Ka-dam-pa faith, it changed hands probably after destruction of the original building. Today it belongs to the Ge-lugs-pa sect and has its head a brother of the Dalai Lama. The main *dukhang* contains the beehive shelves containing the *Kangyur,* or the revealed texts, and the *Tangyur,* or the commentaries on the revealed text, besides the statues of 11 headed Avalokiteswara, Sakyamuni and Tsongkhang (house of the guardian deities) has a large collection of Thangkas. Women are not allowed inside the *gonkhang,* unless the deities are veiled. The Likir monastery also has a school for young novices.

Basgo : Basgo has historically been a centre of events. Basgo castle is a fine reminder of past glory. A natural source of water enabled it to hold out in the event of siege. Basgo village, with its ruined castle, is situated on the right bank of the river Indus, between Nimmu and Saspol. There are several temples attached to the castle, presumably royal temples. One of these, the Maitreya temple is supposed to have been built by Tshewang Namgyal. It has a two-storey high golden statue of Chamba (the future Buddha) which can scarcely be seen whole from downstairs. The murals are specially beautiful, perhaps matched only by those at Alchi.

Phyang : The gompa is situated at the foothills of the Ladakh Range on the left side of the Srinagar-Leh route, about 16 km short of Leh. The Phyang monastery belongs to the Dri-gung Kargyu-pa sect one of the six divisions of the 'Red Hat' order. It has about 50 monks living in its quarters. Its foundation is attributed to Tashi Namgyal, who ascended the throne in the latter part of the sixteenth century. The dukhang is a very old building which seems to have been later renovated. The main altar has statues of Avalokiteswara, Padmasambhava and Kungola-Grangs-pa, the founder of the Drgung-Kargyupa sect. The view from the top of the gompa is spectacular, with the village on one side, while on the other is the deeply furrowed Zanskar range.

Spituk: The Spituk monastery is built on a lone rocky hill on the right bank of the Indus, eight kilometres short of Leh, on the Srinagar-Leh route. It is one of the most influential monasteries of the 'Yellow Hat.' Ge-lugs-pa order has its subsidiary, the Sankar gompa. Built in the fifteenth century, on the ruins of an eleventh century gompa, the Spituk gompa has as its abbot, the Kushak Bakula.

Tsemo: Rising high above the town of Leh is the Namgyal Tsemo peak, studded with historical and religious buildings. Besides the ruins of the fort of Tashi Namgyal there are two gompas, believed to have been built by king Trags-Bumde in 1430. The Maitreya temple at one time had very old murals and inscriptions, but these have since been replaced by more recent ones depicting, as in all 'Red Hat' monasteries, the figures of Avalokiteswara, Padmasambhava and the Sakyamuni. Surprisingly enough, there is also a painting of the founder of the Ge-lugs-pa school, Tsong-Kha pa, and of the Goddess Tara. There is a massive central statue of Maitreya, three storeys high, flanked by smaller ones of Avalokiteswara and Manjusri.

Sankar: Sankar gompa along with the village of the same name, is tucked away deep in a valley behind Leh town, towards the north. It is one of the subordinate monasteries of Spituk, housing about 25 monks who belong to the Ge-lugs-pa order. It is also the residence of the greatly revered Kushak Bakula, who apart from being the head of the Spituk monastery, is also the senior most in the hierarchy of the incarnate Lamas of Ladakh.

Shey : Shey is today a village of no more than a few houses. It was once the capital of Ladakh. Later when the capital moved to Leh, it retained the special honour of continuing to be the birthplace of all royal children. Situated 15 km from Leh, towards Thikse, the ruins of the once glorious palace of Shey stand high up on a sheer rock cliff, and appear to be an architectural extension of the rock formations. Below, at the foot of the rock mountain, are engraved the figures of the five Dhyani Buddhas, each of whom stands on a lotus and can be distinguished by his animal vehicle. They are Amitabha-peacock, Amoghasiddhi-garuda, Vairocana-lion, Akshobhay-elephant and Ratnasambhava-horse.

The Shey oracle is famous for his predictions and undergoes a very strict ritual of discipline and meditation before becoming possessed and going into a trance, after which he is transformed into a fearsome entity, prophesying events and performing cures.

About four hundred metres from the Shey palace is another chapel containing another huge statue of the Sakyamuni. Probably built by Nepalese workmen, it seems to be a more recent replica of the old temple. There are murals of the 16 *Arhats* and the three other preachers—Tsong-Kha-pa, Padmasambhava and Atisa—and many other deities. Among other chapels there is one dedicated to Amitabha (boundless light) and has a printed relief of a thousand Buddhas with different *mudras* (hand gestures). There is a great collection of chortens in Shey scattered over a vast area along the hill range. Perhaps nowhere else in Ladakh can one find such a conglomeration.

Thikse: The Thiske monastery is an impressive complex spilling over a high rocky hill, right down to the Leh-Upshi road. Situated about 17 km from Leh, it is perhaps the most photographed monastery, its numerous monk cells lie scattered, with the red and yellow main building rising majestically high above in beautiful balance. The Thikse monastery belongs to the Ge-lugs-pa sect and houses about 500 monks. Most works housed in the library are rare, some handwritten and are embossed in gold. The famous texts, the *Kangyur* of 108 volumes and the *Tangyur* containing 225 texts, are stacked in the new library in great numbers. Thikse has its own printing unit, apart from a large farm which includes an apple orchard, one of the largest, most serene and attractive Buddha images was installed here in 1984.

Stakna : The Stakna monastery sits like a crown atop a huge rocky hill that appears suddenly amidst lush green fields on the left bank of the Indus, a little upstream of Thikse. It presents a very fine profile from the Leh-Upshi road, though one can approach it more easily from the other side of the Indus. Built like most monasteries of seventeenth century, the Stakna (tiger's nose) monastery belongs to the Drug-pa Kargyud-pa order as does the neighbouring Hemis gompa, and at one time, it enjoyed a greater status than Hemis. The credit for its establishment goes to Sengge Namgyal, and Stag-tsang-pa, though it was Sengge's half-brother Nawang, a celibate who was responsible for its resurrection. As at Spituk, the view from atop this massive crag is breathtaking.

Matho : The Matho (also called Mashro) gompa is the only monastery in Ladakh belonging to the Saskya-pa sect. Apart from the monks, there are many novices attached to the monastery. The *dukhang* is a new building and contains, among other things, several statues of the Buddhas. Next to the *dukhang* is a small chapel which contains some priceless *thangkhas*, unfortunately in a state of neglect. There are also old brocade dresses which the monks wear during the festivals. The *gonkhang* (temple of the guardian deities) is the most interesting of all rooms, since here, as in no other place in Ladakh, the floor has a 50 cm layer of barley grain, an offering to which each family contributes out of its annual harvest. One of the most awe-inspiring events during its principal festival is the performance of monks walking nimbly along parapets, completely blindfolded.

Hemis : Hemis is the most important monastery in terms of spiritual hierarchy as also the richest in Ladakh. It was founded like most other 'Red Hat' gompas, by Stag-tsan-pa under Sengge Namgyal's patronage at the beginning of seventeenth century. Completely hidden from the main route, Hemis is a retreat in every sense. Due to its secluded location it escaped the wrath and robbery of Zorawar's soldiers.

Hemis has gained extensive fame for its two-day festival which, though not very different from the festivals of the other gompas, has the advantage of taking place in the middle of summer.

The *cham* or masked dances are performed by Lamas both young and old, sometimes also by novices barely seven or eight years old. They wear long gowns of heavy brocade and masks, some with serene and benign expressions, while others seem fearsome and grotesque. Animal masks, specially those of deer with large antlers are also worn. There are mystery plays usually dedicated to Padamasambhava, depicting different events of his life.

Hemis possesses a gigantic *thangkha,* believed to be one of the largest in the world. It depicts a beautifully embroidered figure of Padmasambhava, which is displayed for the public every twelve years. There is another *thangkha* large enough to cover the entire wall of the main hall, over which it is draped during the festival.

Tak-Thok : About eight kilometres from Chemre is the village of Sakti lying to the right of the Indus river. The Tak-Thok gompa is situated above the Shakti village, near a cave. It is said that Padmasambhava stopped here to meditate, while on his way to Tibet. It belongs to the oldest order of Buddhism, the Nying-ma-pa.

In the *Lhakhang* (hall of gods), there are images of Padmasambhava and Avalokiteswara, but the wall paintings are covered with black soot and are hardly discernible. The *dukhang,* on the other hand, is a recent addition, with brightly coloured frescoes of awesome divinities and a central image of Padamasambhava.

Chemre : About 45 kilometres from Leh and five kilometres from Karu towards the Chang-Lapass, the Chemre gompa and Chemre village are perched on a high rock. The gompa is supposed to have been built in the seventeenth century by Stag-tsan-pa, under the royal patronage of Sengge Namgyal. After Sengge's death it was completed and dedicated to his memory by his widow. Chemre gompa belongs to the Drug-pa-Kargyud-pa 'Red Hat' order and houses over a hundred monks.

Rangdum : On the Kargil-Padam route, Rangdum is the first village with a monastery, indicating the end of the Muslim belt in the Suru Valley and the beginning of the Buddhist region.

The Rangdum gompa sits atop a lone hill rising in isolation in the midst of a wide valley. Situated at 3,600 metres, this red coloured monastery belongs to the 'Yellow Hat' Ge-lugs-pa order. The *Rimpoche* (incarnate head) is Tensing Chorgyal, younger brother of the fourteenth Dalai Lama; but, since the Rinpoche has relinquished celibacy another Lama serves as the abbot. There are about 50 monks attached to the gompa and many *trawa* (novices) come for their induction into Buddhism.

The *Dukhang* has images of Sakyamuni Buddha, the 11-headed Avalokiteswara and Tsong-Khapa, the founder of Ge-lugs-pa order. The frescoes adorning the walls are of the Mahasidhhas.

Sani : The Sani gompa is situated by the Kargil-Padam route, about 10 km short of Padam. Unlike other monasteries of Zanskar, Sani is situated at ground level and within the village. Belonging to the 'Red Hat' Drug-pa-Kargyud-pa order, it houses about 25 monks. Its most unique possession is a large number of beautiful stucco decorations.

The surrounding consecrated *mane* wall (*mane* means precious stones of jewel) contains figures of *chortens* and deities as well as the *mantra* 'Om manipadme hum' (Glory to the Jewel in the lotus) carved on flat stones in shallow relief.

Just outside the Sani monastery is an old graveyard, supposed to contain the remains of Naropa, the great Buddhist sage from Vikramsila. It has a two-metre high statue of the Maitreya (the Buddha to come), carved out of a single grey rock, apart from smaller statues. In the centre of the yard prayer flags flutter on a tall mast.

Karsha : Karsha is the most important monastery in Zanskar and is situated on the slope of a high mountain to the left of the junction between the Doda and the Tsarap (Lunak) rivers, just above the town of Padam. The approach is from the Tungri Bridge over the Doda river and along its banks.

The Karsha gompa is a 'Yellow Hat' monastery with Tensing Chorgyal, younger brother of the Dalai Lama as the head. However, as in the case of the Rangdum monastery, this gompa is also looked after by a local head Lama.

Phugtal Gompa : Phugtal Gompa (*Phug* means cave), as its name depicts, is situated in a large cave on a precipitous cliff high above the raging waters of Tsarap river. It is a very difficult trek of 2 to 3 days southeast of Padam a little off the Padam-Manali track. The gompa seems to be pasted to the vertical rockface, defying all laws of gravity.

Belonging to the 'Yellow Hat' Ge-lugs-pa sect, this monastery houses between 60 to 70 monks. The cave has a *chorten* believed to contain the relics of the monastery's founder, Tsherap Zangpo, who also founded the monasteries of Rangdum and Likir. There are three prayer rooms in the dark cave with frescoes similar to those of eleventh century Alchi. There is also a slab of dark stone with the inscription, "Csoma de Koros lived here 1825-6 Pioneer of Tibetan Studies", which commemorates the memory of Csoma de Karos, a Hungarian scholar who spent a year in the Phugtal Gompa. He had come to Zanskar to try to seek the origins of the Magyar tongue, and establish its relationship with Tibetan. Though he failed in his primary objective, his extensive studies in the field were duly recognised by the Royal Asiatic Society, Calcutta. To this day he is regarded as the father of modern Tibetology.

We are now in a position to listen to stories of Brigadier Sher Jung Thapa, MVC, Colonel Prithi Chand, MVC, and Colonel Rinchen, double MVC, SM who saved Ladakh. Most of their stories will be recounted in their own words - the way they narrated them, i.e. with minimal editing.

CHAPTER 4

Sher Jung Thapa, MVC, and The Siege of Skardu

11 FEBRUARY TO 14 AUGUST 1948

"........... Throughout the period of siege, Lieutenant Colonel Sher Jung Thapa showed outstanding leadership and great determination in holding out .... inspite of the enemy offering him very liberal surrender terms and knowing that there was no hope of ever being relieved. By his personal example and indomitable spirit he kept the morale of both his troops and civilian refugees at a high level .... his conduct has been in the best traditions of the Indian Army"

\- (Extracts from the citation for Maha Vir Chakra)

Few sieges in the annals of modern war have lasted for 185 days. Before we recount the details of this epic achievement it is worthwhile to take an overview of the setting in which this crucial stand was made. The credit for this heroic deed goes entirely to Brigadier Sher Jung Thapa. Without his stand at Skardu, Ladakh could never have been saved.

Sher Jung Thapa was born at Abbottabad (now in Pakistan) on 18 June 1908. His grandfather and father were both distinguished soldiers. His father - Honorary Captain Arjan Singh Thapa, Sardar Bahadur, Order of British India (OBI) and Indian Distinguished Service Medal (IDSM) was the Subedar Major of the 2/5 Royal Gorkha Rifles for 27 years and six months. Sher Jung Thapa did his college education at Dharamsala. He was an excellent hockey player. As the Captain of his college team he frequently played with the stalwarts of 1 Gorkha Rifles Regimental Centre at Dharamsala. On the hockey field he struck a friendship with one Captain Douglas Gracy who was the Adjutant of the Centre. He encouraged Thapa to join the Forces of Jammu and Kashmir State as an officer and was commissioned on 1 September 1932. Thapa met Gracy again after a lapse of twenty-five years in different circumstances. Lt. Col. Thapa was a prisoner of war in Pakistan and General Sir Douglas Gracy was the Commander-in-Chief of Pakistan Army. More details of this meeting later.

In the earlier chapters we discussed very broadly how the British policy of divide and rule culminated in the tragedy of the partition of India. Millions of people had to leave their homes and hearths. Almost four million people were butchered in the communal frenzy which was triggered off as a chain-reaction to the direct action launched by the Muslim League in Calcutta in 1946 to establish a Muslim homeland. We also discussed how, in event, the Muslim community, which was used by the British to bring about this tragedy had been the worst sufferer of the political and economic consequences of the partition.

The Prime Minister of Jammu and Kashmir, Ram Chandra Kak, had assured Pakistani politicians that he would be able to persuade the Maharaja to accede to Pakistan. The Maharaja who dreamt of making Kashmir into a neutral Switzerland of Asia did not take kindly to Kak's activities and dismissed him. To coerce the Maharaja to accede to Pakistan that country imposed a blockade of the State despite the standstill agreement. Vehicles carrying essential supplies for the forthcoming winter were prevented from entering the State. However, the final act which precipitated the accession of the State to India was the launching of the Tribal Invasion on 20 October, 1947. It left the Maharaja with no option except to accede to India and get military aid to save his State. This invasion has been aptly described by Mr. Khurshid, the Kashmiri Private Secretary of Jinnah, as a "criminal folly which sealed the fate of Kashmir". Khurshid a bright political activist was picked up by Jinnah a couple of years earlier as his Personal Secretary. He was sent to Kashmir to organise those Muslims who were in favour of joining Pakistan. He had worked for many weeks to organise those who were not in jail. The Maharaja had imprisoned most of the political leaders of all parties when in 1946 the National Conference, with the support of Nehru, had launched a Quit Kashmir movement against the Maharaja. Khurshid was essentially working on the cadres of Muslim Conference who had assured Ramchandra Kak that Pakistan will permit the Maharaja to continue to rule Jammu and Kashmir if he acceded to Pakistan.

We also discussed the role of Lieutenant Colonel Roger Bacon, the British political agent at Gilgit. He, obviously, masterminded the coup d'etat which Major Brown would stage by issuing the code word "Datta Khel" in case the Maharaja acceded to India. Datta Khel, incidentally is a well known centre of Tochi Scouts towards South West of Peshawar, in the heart of the region inhabited by the Tochi Tribe. It will be recalled that Major Brown had earlier served with the Tochi Scouts at Datta Khel which was not too far from the new location of Lieutenant Colonel Roger Bacon who moved from Gilgit to become the political agent in

Khyber Agency. Brown was retained by the State as the Commandant of the Gilgit Scouts. Brigadier Ghansara Singh had protested against retaining any British Officer in Gilgit when he was nominated as the Governor, but Ramchandra Kak paid no heed to it.

When in July 1947, Brigadier Ghansara Singh was nominated to go as the Governor of Gilgit, he was working as the Brigadier-General-Staff of the State Army and was well posted with intelligence reports. It is relevant to record his assessment of the political conditions in Gilgit where he was to go.

Gilgit was made up of six-handkerchief sized principalities. The six rulers were divided into two rival groups. One group was inclined towards Pakistan, while the other wanted to continue their traditional links with the Maharaja. The one leaning towards Pakistan was led by the Mir of Hunza and the other was led by the Raja of Punial. The grouping was :

Mir of Hunza	Raja of Punial
Mir of Nagar	Governor of Koh Khizar
Governor of Yassin	Governor of Ashkomand

After the termination of the lease of the Gilgit Agency to the British on 3rd June, 1947, the Mirs of Hunza and Nagar were State guests at Srinagar during the month of July. Ghansara Singh knew that a political activist from Poonch, Abrahim, was escorted by Captain Ihsan Ali to the State guest house for mid-night meetings with the Mirs, obviously to persuade them in favour of Pakistan.

When they returned to Gilgit in August, they were, as per the State protocol, entertained by Ghansara Singh. The Mir of Hunza showed the Pakistan flag which he was carrying, to some of his friends at Gilgit. This news spread in Gilgit town and reached the Mir also. He wrote a private letter to Ghansara Singh that this was a false canard spread by his enemies.

Subedar Major Babar Khan of the battalion-sized Gilgit Scouts was married to the sister of the Mir of Hunza. He was also the uncle of the Mir of Nagar. Among all the Frontier Scouts, the Gilgit Scouts, because of their role against the Soviet Union and China, were an elite outfit and trained with dedication to march upto sixty miles a day.

We also discussed the geography of the region in which Gilgit, Skardu, Kargil and Leh lie. It lies on the northern side of the Himalayas. It is a continuation of the Tibetan plateau. The Indus river which has its source from the holiest and highest major lake in the world, Mansarovar, runs through this region from east to west. The main routes, mere pony tracks, linking the four major towns ran on the southern and northern side of the Indus river. Administratively, this region was called Ladakh, except Gilgit which was on lease to the British. Leh (11,000 feet) was the summer capital of Ladakh while the administration shifted to Skardu (7,500 feet) during the winter.

The State Government deployed one Battalion in the region spread over from Bunji on the border between Ladakh and Gilgit and the town of Leh. During late summer 1947, 6th Jammu and Kashmir Infantry (State Force) was inducted into this region to relieve the 5th Battalion as a routine turn-over of the units.

The 6th Battalion was composed of two companies of Muslims and two companies of Sikhs with a sprinkling of Dogras. The Battalion was commanded by Lieutenant Colonel Abdul Majid Khan. Most of the Muslim soldiers were from Poonch region. Poonch Muslims, as we discussed earlier, were fully infected with the communal frenzy which was gripping much of Northern India. When the leading company composed of Muslims and commanded by Capt. Hussain Khan was moving in the vicinity of Bandipur, the men shouted pro-Pakistan slogans. Reports of this did reach the State Army Headquarters at Srinagar. The commanding officer Lieut. Colonel Abdul

Majid Khan was asked to enquire. Being 'I don't want trouble' type of a man, he closed the matter by recording that the men had shouted only religious slogans. Brigadier Ghansara Singh's request not to send this Battalion to Gilgit was ignored.

By the month of October, Brigadier Ghansara Singh had fully assessed the developing situation. He was unaware of the planned coup, but got enough information from some elements loyal to the State to be able to write to the Prime Minister and Deputy Prime Minister about the situation in Gilgit. In a nutshell, he warned them that if the Maharaja acceded to India, the Gilgit province would go to Pakistan because the British officers of the Gilgit Scouts as well as Muslim officers in the 6 Jammu and Kashmir Infantry wanted Gilgit to go to Pakistan. With trouble going on in the Jammu and Poonch region there was nothing that the State administration could do to help the Governor of Gilgit; the six battalions of the State Army, its total Force, were fully committed.

Ghansara Singh also started getting reports that Forces were being organised in the States of Swat and Chitral to invade Gilgit. These States which bordered on Gilgit had already acceded to Pakistan after the partition on 15 August, 1947.

6th Battalion of Jammu and Kashmir Infantry as a part of the Kashmir Brigade of the State Forces was deployed as under :

Under Operational Control of Major Sher Jung Thapa, the Second in Command: HQ at Leh

Leh - Company minus Sikhs

Kargil - Two platoon mobile column of Sikhs, commanded by Captain Ganga Singh

Skardu - Company minus Muslims, commanded by Captain Nek Alam Khan.

Under Operational Control of Lieutenant Colonel Abdul Majid Khan, the Commanding Officer: HQ at Bunji (34 miles from Gilgit).

Bunji - One Company of Muslims, under Hussain Khan. One company minus Sikhs.

The Sikh company of Bunji was made up of fresh recruits who were not yet fully trained soldiers as they had not so far fired their weapons on a classification range. So the Commanding Officer reported to Brigadier Ghansara Singh that at Bunji, he had only one company fit for an operational role. 24 October, 1947 was the Maharaja's birthday. The Tribal Invasion had started at the night 20/21 October and reports of advancement towards Uri and Baramula started reaching Srinagar. The Government was still confident of countering the invasion and in order not to cause panic, the functions to celebrate the Maharaja's birthday were held as usual.

At Gilgit, all the officers of the 6th Battalion were called for the birthday celebrations. Ghansara Singh learnt later, after he was arrested, that this meeting between the Muslim officers of the Battalion with Major Brown and other officers of the Scouts provided an opportunity for finalizing the planned coup.

On the intervening night of 31 October/1 November, code word 'Datta Khel' was issued. Ghansara Singh, his batman and driver fought with their personal weapons but eventually the contingent of Scouts led by Subedar Major Babar Khan put the Governor under arrest.

Captain Matheison with a contingent of the Scouts and the Muslim company of 6 Jammu and Kashmir Infantry at Bunji surrounded the Sikh company and after a few days of resistance they were captured. We have already narrated (in Chapter 2) how Major Brown sent two telegrams on 1 November to Khan Abdul Qayyum Khan and Lieutenant Colonel Roger Bacon about his success.

The next three months at Gilgit were spent in organising a force for the capture of Ladakh. Soldiers from Chitral and Swat reached there. Two thousand men were recruited from Hunza and Nagar. Weapons and ammunition were para-dropped. Brigadier Ghansara Singh who was held a prisoner at Gilgit recalls, "I could see the para-drops" and information continued to reach him through elements in the Scouts still loyal to him.

Towards the end of January Lieutenant Colonel Roger Bacon flew into Gilgit. He met the civil and military officials and then flew back to Peshawar. This was the green signal to launch operations to capture Ladakh.

On the Indian side, after the first landing of troops at Srinagar airfield on 27 October the build-up continued. It was mostly by air because the mud road from Pathankot to Jammu, with no bridge on river Ravi, was hardly suitable for military traffic. Slowly, but surely, the invaders were pushed back from the outskirts of Srinagar, and by December Baramula had been recaptured. Till then no one had any time to pay attention to what was happening in the trans-Himalayan region of the State. The problem came to life when the Gilgit coup d'etat by Major Brown was reported, and by about the middle of November the Muslim elements of the 6th Battalion of Kashmir Infantry alongwith the Gilgit Scouts had established firm control over Gilgit. As we have discussed earlier, it was in the beginning of December that Major Prithi Chand reached Srinagar with 2 Dogras under the command of Lieut. Col. Gopal Gurunath Bewoor (later to become the Army Chief). When Prithi Chand started agitating to go to Leh the problem came into some focus.

Kashmir Brigade at Srinagar, had in the meanwhile, ordered Major Sher Jung Thapa at Leh to move to Skardu with as many troops from Leh and Kargil as could be spared. He left Leh on 23 November 1947 and reached Skardu on 3 December with a contingent of two Officers, two Junior Commissioned Officers and 75 Jawans who were mostly Sikhs. During his move Thapa learnt that Pakistani agents were active throughout Ladakh, but had not made much headway in subverting the people except in the town of Skardu. He was given the rank of Lieutenant Colonel and ordered to defend Skardu. It is best that from here onwards we let the old war horse narrate the story in his own words.

LIEUTENANT COLONEL SHER JUNG THAPA'S STAND AT SKARDU

With our arrival, the scene of activity shifted to Skardu. Actually the pot was already boiling but the lid was not yet off. On our way to Skardu on 2 December 1947 between Parkuta and Gol we sighted an aircraft (Havard) which was probably on a reconnaissance mission. It circled low over us and then flew off towards Gilgit. Later it was found that it was a Pakistani aircraft.

Skardu is a small and narrow valley at a height of 7500 feet above sea level. This valley is divided into two parts by the river Indus. The area on the northern side of the river is fairly flat, about 12 miles long with its width varying from 1 mile to 4 miles, but very sandy. There is only one long hill called Lukro. This area is totally uninhabited. River Shigar passes through this area and joins River Indus. The area south of the river is fairly flat, about 12 miles long with a maximum width of 5 miles near Skardu. The width decreases as one goes north-west terminating at village Hoto. Along the river bank there is a sandy belt about one mile wide. Except for this sandy waste the whole valley is very fertile and thickly populated. In Skardu town all offices of civil administration, hospital, school and the military post were located.

At Skardu there is a lone hill with its peak at 8853 feet which commands practically the whole of the flat ground described above. The base of the hill is an elipse whose major and minor axes are approximately one mile and half a mile respectively. The south-eastern slopes of this hill are too steep for climbing and the peak is accessible only from the south-west and north. About half way up the eastern slope is a small fort which was not visible from point 8853 because of an intervening hump.

The valley is surrounded by high, steep and barren mountains. Five prominent heights which command the whole valley are :-

Marshkala	16905 feet
Shimshak	18403 feet
Thurgo	17246 feet
Burzila	15800 feet
Thalanka	16830 feet

River Indus is fordable by raft practically at all points in the valley. It snows heavily during the winter and the whole valley remains covered with approximately three feet of snow till about the middle of April. Routes to Skardu from Gilgit and Kargil along river Indus remain open throughout the year (though difficult in winter) where as all other routes including that from Srinagar remain closed during winter.

GATHERING STORM OVER SKARDU

Before our arrival at Skardu, the Wazir (Deputy Collector) Shri Amar Nath Mahajan had seen the signs of the gathering storm. Tehsil Skardu had a much larger area than a normal Tehsil. There were five Jagirdars called Rajas, who exercised considerable influence over the population in their respective areas. They were Raja of Rondu; Raja of Khapalu; Raja of Shigar; Raja of Skardu and Raja of Kharmang.

Wazir Amar Nath had called a conference of these Rajas to obtain political information about their areas and also to sound them about their reactions to the fast changing pattern of political life in and around Skardu. He also wanted to ascertain to what extent these gentlemen would stand by him and help him in procuring rations for the increasing number of State troops. The Raja of Rondu did not attend this conference. The Wazir was not surprised. For, by then reports had been flowing in about Rondus' activities. He was said to be maintaining constant contact with Gilgit Scouts and was urging them to strike at Skardu immediately. It was his entreaties that brought the enemy on to Skardu earlier than anticipated. Later on, it was also learnt that Rondu had been intercepting our civilian couriers and agents operating in Stak (NB 9813) area. He threatened them with extermination. The agents had to lie low and when the threats became very serious, they withdrew.

DARK HORIZON

Capt. Nek Alam of our 'C' Company had intercepted and captured two letters from Captain Hassan Khan of 6 Jammu and Kashmir Infantry. Hassan Khan was inciting the Muslim elements among our troops to revolt and seize Skardu and to take over the administration till the Pakistani masters reached Skardu from Gilgit. These letters were sent by Captain Nek Alam to Jammu and Kashmir State Forces Headquarters at Srinagar. On 4 December four Sikh survivors from Bunji Garrison were brought in by Captain Nek Alam. On receiving information from locals he had gone to village Stak to rescue the party. He could, however, only rescue four wounded and sick, the rest having been taken away towards the Gilgit side by the enemy.

Soon after my arrival in Skardu, I went forward to Rondu to reconnoitre the area. I had thought over the probable routes of approach by the enemy. In winter, attack on Skardu from either Bunji or Gilgit could only be launched via Rondu. All other approaches become impossible due to heavy snow.

After my initial reconnaissance I decided to put one platoon on either side of River Indus at a place called Tsari about 20 miles North West of Skardu to act as outposts. Accordingly, one Muslim platoon under Captain Nek Alam was posted on the right bank and a Sikh platoon under Captain Krishan Singh on the left. My plans were, that on receiving sufficient reinforcements from Srinagar, I would give first battle to the enemy at this tactical point which was a bottleneck. The reason why I could not put more troops immediately at Tsari was the paucity of troops at Skardu for local protection of the town and the treasury. I knew that without sufficient reinforcements and necessary supporting weapons, I could not do much in the event of a planned attack by the enemy. For this reason I was constantly asking Kashmir Brigade to rush reinforcements to Skardu. Consequently, the first batch of reinforcement (strength 90) under Captain Prabhat Singh arrived in Skardu on 10 February, 1948. I had constructed a rough landing strip of about 1000 yards in area Ranga in the hope of receiving air supply.

DIFFICULTY OF REINFORCEMENT

Trekking along the mountainous terrain in snow and blizzards was indeed a Herculean task for our reinforcement which had started from Srinagar on 15 January, 1948 moving in small parties of platoons. This was perhaps the first batch of Army to cross Zojila in the month of January against overwhelming odds and hostile weather conditions.

Suddenly, on the morning of 11 February, 1948 at dawn, Skardu was attacked and surrounded by the enemy (about 600 strong) compared to our own strength of 130. Enemy opened up with all he had, 3 inch and 2 inch Mortar, MMG, LMG and rifles fire. The local sympathizers raised loud battle cries and the effect of surprise and the din would have unnerved less determined troops than my garrison boys. We hurriedly took positions and fought for six hours. The enemy was routed. They left 10 dead, one prisoner-of-war and useful booty of arms and ammunition as under :-

MMG	-	One
Rifles	-	Four
2-inch Mortar	-	One
303 Ammunition	-	Two boxes
3-inch Mortar Ammunition	-	Two boxes
2-inch Mortar Ammunition	-	One box

Seven of our gallant boys laid down their lives and 16 were wounded (4 seriously)

The enemy was led by Major Ihsan Ali, a deserter from 6 Jammu and Kashmir Infantry. Other officers with him were Captain Mohamad Khan - also a deserter from 6 Jammu and Kashmir Infantry - and a Scouts Officer Lieutenant Babar Khan.

While the battle was in progress it was found that the Muslim elements totalling 31 who were in Skardu were missing. Apparently they were informed about the impending attack, through secret agents and deserted during the night stealthily in order to join the enemy.

The enemy killed some of the non-Muslim civilians including the Wazir, Shri Amar Nath, Engineer B.L. Gupta, a Sub-Inspector of Police and the Headmaster of a High School. They also looted the Tehsil treasury which was about a mile and a half from the fort. I had suggested to Amar Nath to move the treasury into the fort but he declined.

The three Muslim wireless operators had also deserted. The set was in one of the rooms of my bungalow. I operated it during the battle in order to pass information back to higher authorities. My having been once a Brigade Signal Officer stood me in good stead while operating the set.

Before attacking Skardu, the enemy had completely annihilated the Sikh platoon under Captain Krishan Singh. The Muslim platoon under Captain Nek Alam had joined hands with the enemy. It was later confirmed that some of the Sikh Jawans were killed in action, while the others including Captain Krishan Singh were murdered in cold blood after being taken prisoners. Tsari was attacked at night 9/10 February 1948.

The fanatic Muslims in the area were helping Pakistani forces so completely that not a word was uttered by them about the attack on Tsari and the massacre of the platoon. Every Muslim civil officer (Tehsildar, Naib Tehsildar etc.), school boys, servants of local shop-keepers and all who were employed to secure information about the enemy did not do so. They knew that the enemy was coming. They took all precautions to see that we were kept in the dark and then were attacked by surprise. This attitude of the Muslims there goes to prove how determined, united and eager they were. They wanted the Pakistan flag to fly over them. The Pakistan agents here did a thorough job.

Fanaticism had made them securityminded and to them "Pakistan" was the sweetest word.

Outpost at Tsari had communications with Skardu only through local runners and occasional patrolling from Skardu. The last I heard from Tsari was on 9 February 1948 to say that all was well. On 11 February 1948, a ration party was scheduled to go to Tsari. from Skardu. The attack on Skardu on 11 February was launched before they could leave.

One more batch of reinforcement, about 70 strong, arrived on 13 February under Captain Ajit Singh followed by yet another of the same size on 15 February Surely, the enemy knew that our reinforcement were on their way to Skardu. Any other enemy commander in place of Major Ihsan Ali Khan would have paid more attention towards effective interception of the move of these reinforcements and would have prevented them from arriving at Skardu. He merely contented himself by erecting some simple road blocks in the way of our troops which were no obstacles at all.

By this time the Skardu Garrison grew to about 285. With this meagre strength, organisation of defence of Skardu was a difficult task. Due to administrative difficulties, lack of sufficient troops and the hostile attitude of civilians, the hill feature (Point 8853 which was so essential for our local safety), could not be occupied. Much against our will we had to contract the ring of our defences.

My pressing demands to authorities in Srinagar to rush more reinforcements produced little results. To make up the minimum strength necessary to man all defences, I had to organise the non-Muslim young civilians who could be usefully employed as riflemen or porters to carry rations and water to the posts. All of them did their best to relieve the Jawans to perform picquet duty.

The enemy activity had turned many local people into penniless refugees. As already mentioned, all the civilian officers were brutally killed by the enemy. We shouldered cheerfully our sacred duty to look after, protect and feed these unfortunate people. All non-Muslim refugees were brought into the Fort and the Muslim refugees who sought shelter were kept outside the Fort, but inside the perimeter. These numbers were :

Non-Muslim refugees	-	229
Muslim refugees	-	19
Muslim Civil prisoners	-	22

THE PICTURE

The enemy occupied Point 8853 on night 14/15 February and started sniping from there. The defences were organised into picquets as per sketch 3. No major engagement with the enemy took place until 24 February, 1948 when one of our fighting patrols, of a platoon strength, encountered an enemy ambush laid by about two platoons. The encounter lasted upto 1800 hrs in which two of our soldiers were seriously wounded.

Our activities were mainly restricted to intensive patrolling within a radius of 5 miles in all directions, in order to keep the enemy as far as possible, to gain time for more reinforcements from Srinagar to join us. The enemy, on the other hand, started bringing in more reinforcements with a view to lay siege of the Garrison.

On 4 March, 1948 from 1600 hrs the enemy shelled three of our picquets namely School, Raja and No. 6 Picquet heavily with 3-inch and 2-inch Mortars. The only casualty on our side was one wounded seriously. We replied to the mortar fire with a few 3-inch mortar bombs which silenced the enemy.

On 6 March enemy assaulted one of our picquets with two platoons under cover of 3-inch and 2-inch mortar fire.

Enemy also used rifle grenades. The attack was repulsed by our troops aided by our 3-inch mortar neutralizing the enemy supporting fire. The enemy left behind one dead. We suffered no casualty.

All the time the enemy strength, both in men and fire power, went on steadily increasing : the intensity and frequency of attacks correspondingly increased. Our patrols began having more frequent encounters with the enemy patrols. On 24 March an enemy fighting patrol walked into our ambush at about 1600 hours and suffered heavy casualties, but on account of heavy fire from enemy MMG and LMG from Point 8853 we could not secure the enemy dead bodies.

On 15 March for the first time some enemy detachments were seen moving towards Go1. Similarly, on 16 March a bigger column of about 2 companies was again seen moving from the direction of Village Kuardo towards Go1. We judged that the enemy's intention was to intercept the column commanded by Brigadier Faqir Singh and assisted by Major Coutts which was coming towards Skardu. The enemy move was immediately reported to Sri Division as a good target for an air-strike, but alas, in vain. As feared, however, on 17 March our reinforcement was completely ambushed and routed by the enemy at Thurgon Pari about 10 miles short of Skardu and historically notorious for ambushes.

According to the information with me, based on a signal message from Sri Division, the column was scheduled to reach Skardu on 18 March but unluckily the column arrived on 17 March, a day too early, only to meet disaster. On this day troops at Skardu were subjected to heavy mortar and MMG fire to prevent us from going to meet the column. However, acting on the message, I had pulled out some troops from the picquets and had moved

Kargil in December 1948 after it was recaptured by 77 Para Brigade advancing from Zojila

Kharal Bridge near Kargil destroyed by the Pakistani invaders while retreating in December 1948.

out towards Go1, along with Captain Ajit Singh and two weak platoons to receive the column. But, the locals told us that the column had already been beaten back by the enemy on the 17th.

This tragedy and the consequent disaster could have been averted if the column commander had a radio set with him to keep in constant touch with the destination and the base. Had the column arrived on the 18th as scheduled, the enemy would have been in trouble with me on his one side and Brigadier Faqir Singh on the other.

During our advance to Go1 we escaped a serious enemy ambush and had to withdraw for about 5 miles, having to fight the enemy on three directions. The battle ended after two hours with no casualties to us. The enemy was, to our misfortune, augmented with bountiful booty abandoned by Brigadier Faqir Singh's force, consisting of arms, ammunition, equipment and clothing. This must have considerably raised his morale.

After the defeat of this column which made a substantial addition to the enemy resources, I began to worry seriously about my own rations, ammunition and man-power which were decreasing. I began to fear that, with the present stock of rations and ammunition I would have to carry on for a much longer period than previously anticipated.

The ration stocks were meant only for 80 men upto the end of August, but now the strength had increased ten fold (including refugees). Efforts were made to augment our stocks by requisitioning stocks from the civil bazar and from the vacated houses of the local population. Civilian refugees helped me much in collecting rations, sometimes under the fire of the enemy. But, even then, our stock was far from adequate and I was compelled to reduce the ration of the fighting men, which ultimately went down to 180 grammes of atta a day with 30 grammes of dal and no ghee at all.

In the face of such a danger, I further repeated my request and urged the authorities to rush fresh reinforcements without any loss of time. I also imposed strict control on the use of ammunition. We had only two hundred mortar bombs for 3-inch mortar, at the time of siege. To conserve these bombs, I used to sit with the mortar crew and allowed only 2 or 3 bombs to be fired when needed. The crew consisted of Dogra and Sikh soldiers. At the time of firing they would shout 'Durga Mata Ki Jai' (glory to Goddess Durga) and 'Jo Bole So Nihal - Sat Sri Akal' (whosoever prays to true God is rewarded). It was strange that each time we fired we squarely hit the target and inflicted casualties. As a result, at the time of surrender, we still had 80 mortar bombs left and I had to destroy them along with other arms and equipment; we had no other ammunition left at all.

As a result of my requests, advanced elements of a force named Biscuit Column were reported to have arrived at Parkutta on 31 March. This column consisted of elements of 4 Jammu and Kashmir Infantry plus one company of 7 Jammu and Kashmir Infantry, and a mortar platoon under the command of Lieut. Colonel Sampuran Bachan Singh, assisted by Major Coutts. I later came to know that Lieut. Colonel Sampuran Bachan Singh had been replaced by Lieut. Colonel Kripal Singh of 5 Jammu and Kashmir Infantry.

This column advancing by companies was held up and harassed by the enemy at various stages. Only three companies reached Parkutta, but could not progress any further. Lieut. Colonel Sampuran Bachan Singh and Major Coutts both escaped to Leh along with two radio sets, thereby leaving no means of communication with the advanced troops at Parkutta. The absence of radio communication must have caused great difficulty and adversely affected the morale of troops.

I recently read with interest the following account of the Biscuit Column in the official history :

"Meanwhile, Lieut. Colonel Thapa had been asking urgently for more reinforcements, and his demands were obviously justified. Accordingly, about one company of the State Force troops with two 2-inch mortars and two MMGs had been dispatched from Srinagar on 17 February for Skardu under the code name of "Biscuit Column". A platoon of troops joined the column at Kargil, and Brigadier Faqir Singh, the column commander, moved from there on 8 March for Parkutta and Gol. He was accompanied by Major Coutts, an officer of 6 Rajputana Rifles of the Indian Army. The cold was intense and the men suffered terribly, but Brigadier Faqir Singh consoled himself with the thought that the atrocious weather would have immobilized the enemy also. He was wrong. The hostiles knew of his coming and had made their plans to deal with him. They pulled out most of their troops from the siege of Skardu, while those left behind kept up a smart fire on Skardu fort to deceive the garrison. About 100 of the raiders set up an ambush near Gol, about 32 km east of Skardu, where the track went through a narrow gorge. They were outnumbered by Brigadier Faqir Singh's force, but the ground favoured them, the element of surprise was on their side, and ambushing was the game they all thoroughly understood. It was 15 March when Lieut. Colonel Thapa's look-outs saw some raiders marching towards Gol. The next day, a bigger body of the raiders was seen hurrying towards Kuardo and Gol. Lieut. Colonel Thapa had been informed that a relieving column under Brigadier Faqir Singh was moving upto Skardu and was expected to reach him on 18 March. So, he immediately realised that the hostiles were going to intercept the column before it reached Skardu. But, he could not warn Brigadier Faqir

Singh, whose single W/T set had broken down at Kargil. Thapa, however, reported the hostile movement to Srinagar, suggesting an air strike against the raiders. Unfortunately, the Air Force was fully pre-occupied elsewhere, and no air strikes took place."

"On 17 March, Brigadier Faqir Singh's column entered the gloomy gorge on Thurgo, midway between Gol and Skardu hoping to reach Skardu before long. The raiders were nowhere in sight. But, suddenly bullets began to rain down on it from the steep heights on either side. The startled troops tried to rally and hold on, but soon 3-inch and 2-inch mortar bombs came crashing among them and inflicted several casualties. The Brigadier was himself wounded, and panic seems to have prevailed and in the confusion, the raiders succeeded in looting the baggage train and the ammunition boxes. The column fell back, having lost 26 men killed, 7 missing believed killed and 18 wounded. Much booty and war material fell into the hands of the raiders, including 27 rifles, five sten guns, two Vickers Berthier medium machine guns, 64,000 rounds of rifle ammunition, 84 hand grenades, 51 HE bombs of 2-inch mortar and 204 HE bombs of 3 inch- mortars. The column reassembled at Kargil. By 2 April, Brigadier Faqir Singh was back at Srinagar leaving his troops under Major Coutts at Kargil. Thus ended the second attempt to reinforce Skardu."

Meanwhile, at Skardu, the enemy started full scale operations. Shelling of our positions became more frequent accompanied by occasional attacks. We, in our effort to conserve our resources, had to keep to minimum firing in reply. On one occasion, the enemy fired no less than two hundred 3-inch mortar shells on two picquets. Our casualties that day were two killed and four wounded.

The enemy began to gain more ground and thus tightened the ring around us, which compelled us to reduce the extent of our patrolling activity. The lack of sufficient food, coupled with operational pressure began to tell adversely upon the health of the troops. The hospital cases too increased both from sickness and battle casualties. The medical equipment and medicines were not enough to cope with this development. The wounded and sick who just could sit were employed to man the defence in the biting cold. These men, though their vitality was greatly reduced, accepted these tasks cheerfully, fully realising the helplessness of our situation.

The tempo was now rising. On 28 March at 0300 hours the enemy launched a well prepared and determined attack with two platoons led by cattle on No.6 picquet held by two sections. The attack was supported with fire from 3-inch and 2-inch mortars and MMGs from Point 8853. Our troops fought gallantly and repulsed the attack after two hours battle.

Two hours later i.e. at 0500 hours the enemy launched yet another attack, this time on a much larger scale on all our picquets and from all directions. A furious battle developed which lasted throughout the day. At night, however, enemy withdrew from all posts except two viz. School and Raja picquets commanded by Captain Ajit Singh and Jemadar Piara Singh respectively. They substantially increased pressure against these posts. The gallant warriors in these picquets rose to the occasion and stood up well. Next morning at dawn we successfully counter-attacked with a platoon and the enemy fell back, leaving behind 20 dead which we actually counted. We suffered four killed and ten wounded.

After this punishment the enemy realised that inspite of their superiority in number and weapons, they would be required to pay a very heavy price if they persisted in their attempts to break through those defences made impregnable by the fearless spirit of the besieged fighters. What added strength to our defences was a carefully thought out fire plan and the doubling and trebling of the roofs of bunkers and posts which made them practically immune against 3 inch-mortar shells.

Another similar bid to break through our defence on 17 April was again beaten back imposing on the enemy a caution never to embark on such an enterprise. On 10 April I sent the following signal to the Division.

"Hostile pressure terrific. Our key picquet cut off last four days and fighting without water. Its fall certain today and fort cannot survive one hour after that. Only immediate air support can save us now".

However, there was much fight still in my troops. A fierce-counter attack, again by a platoon next morning led by Naik Chatru of 6 Jammu and Kashmir Infantry cut its way through the enemy and drove the Pakistanis away. Three dead from the enemy were later found in one of the houses cleared by us. Most of the dead and wounded enemy soldiers had been evacuated earlier.

The enemy, however, had succeeded in establishing his positions very near to ours - in some cases only 20 yards from our position. It was an eyeball to eyeball confrontation. I was not in the full picture as to what the higher authorities were doing to relieve me.

The following extract from the official history tells the story and fills the gap in Lt.Col. Sher Jung Thapa's story :

Reinforcements were already under way for Skardu. The Indian Army units were fully occupied in the bitter fighting of Uri and Poonch sectors. But, all available troops of 5 Jammu and Kashmir Infantry and 7 Jammu and Kashmir Infantry, about one and a half battalions, were collected for the relief of Skardu. On 30 March, seven officers and about 400 men of 5 Kashmir Infantry reached Srinagar from Jammu. On 1 April, a party consisting of one officer, one VCO and 40 other ranks was reaching

Kargil, and another party of three VCOs and 30 other ranks was only one stage behind it. With the help of these reinforcements, Major Coutts was ordered by 163 Brigade to make Kargil into a firm base, defended by a company of troops, a detachment of mortars and a section of MMGs. He was also instructed to move one company forward to Bagicha and another to Parkutta, converting each of these localities into defended bases.

On 3 April, the first batch of 5 Kashmir Infantry left Srinagar for Kargil. It consisted of one officer and 37 other combatants and carried a 3-inch mortar, a 2-inch mortar, 3 Brenguns and plenty of ammunition. Lieut Colonel Sampuran Bachan Singh of the Indian Army, who had been till then a Major and 21C of one Sikh, was commander of this batch, and his column was named the 'Sugar Column'. His instructions were to relieve Skardu, leave there the majority of his troops to strengthen the garrison, and return with the civilians there. For this purpose, he was made the commander of all the State Forces in the Kargil-Skardu area. He was also informed that two companies of seven Kashmir Infantry and a section of MMGs were already at Kargil under Major Coutts, and that one company of seven Kashmir Infantry with a section of 3-inch mortars was just then reaching Kargil. Major Coutts was to act as Staff Officer to Lieut. Colonel Bachan Singh.

Another batch of five Kashmir Infantry followed Lieut. Colonel Sampuran Bachan Singh the next day. It consisted of 70 officers and men and carried two 2-inch mortars, 6 Brenguns, 14 stenguns and plenty of ammunition. On 8 April the third batch of five Kashmir Infantry left Srinagar under the command of Lieutenant Colonel Kripal Singh of the five Kashmir Infantry. His force consisted of 69 men and carried one 2-inch mortar, 3 Brenguns and 9 stenguns. He was designated the commander of the relief column to Skardu. Lieut. Colonel Sampuran Bachan Singh was to be left behind between Kargil and Skardu to hold one of the strongly defended localities.

There were now about two battalions of the State Force moving on to Skardu. It was; for the situation, a strong force as shown by its armaments listed below :

Item	**Biscuit Column**	**Sugar Column**
Rifles	255	252
Stenguns	48	61
Brenguns	27	23
Pistol Signal	6	10
3-inch Mortars	9	7
2-inch Mortars	2	3
MMGs	2	4

Lieut Colonel Sher Jung Thapa was sending urgent requests from Skardu for early reinforcements. On 11 April, therefore, 163 Brigade ordered Major Coutts to advance immediately from Kargil towards Skardu. From this date onwards, there was a stream of men all the way from Srinagar to Kargil moving up towards Skardu. On 14 April, when Major Coutts' party was 14 km beyond Bagicha, the first batch from Srinagar was just entering Kargil, the second batch was near Dras, the third at Matayan, and the fourth and fifth at Gumri. The sixth batch was just leaving Srinagar.

By then, the troubles of these columns were growing. On 17 April, Major Coutts reached Parkutta after a long night march. He found the village silent and deserted. He reported back that further advance was impossible unless 100 pack ponies were sent to him from Kargil. He also wanted five signalling lamps and a dozen W/T sets of type No. 48, because it was not possible with the single No. 22 W/T set with him to co-ordinate the advancement of his column, which had to move along the river bank for the sake of tactical support. Major Coutts also complained against the weather, which was cold and wet. Two days later, he reported that the raiders had surrounded his column, and were sniping it from the heights. Supplies of ammunition were wanted immediately and he concluded with the hope that though his troops were raw, they "definitely should hold" the attacks. To this, Headquarters 163 Brigade replied that his troops were certainly far better than the irregulars opposing him, and he should explain how his stock of ammunition was expended in a single short engagement.

The next day, 20 April, Coutts informed 163 Brigade that his force lost 6 men killed and 9 wounded the previous day, and that 2 light machine guns and 3 rifles had fallen into the enemy's hands. The raiders were two platoons in strength, with one 3-inch mortar, two 2-inch mortars, two medium machine guns and 6 Brenguns. They were well trained and had the fullest support of the locals. His own troops had low morale, the ground was unfavourable to him, and he had to leave behind strong garrisons at the main villages as he advanced. Due to the shortage of transport, only a moderate amount of ammunition had been carried by his column. Finally, Major Coutts requested immediate air attacks on the raiders opposing him.

The hostiles kept up their desultory firing on Coutts' force and, on 21 April, tried to infiltrate into his position via the Katicho Nala. The attempt was beaten back. But, the raiders kept up their pressure and, on 24 April, Coutts reported that all his troops were already committed in the fighting and he had no reserves left. He urgently wanted reinforcements, ammunition and rations from Kargil, where Lieut. Colonel Sampuran Bachan Singh had arrived with his force on 21 April.

The day after reaching Kargil, Lieut. Colonel Sampuran Bachan Singh received orders from HQ 163 Brigade to push on to Parkutta and to carry with him rations, ammunition and No. 48 W/T sets for Coutts. On 26 April, the Sugar Column under S.B. Singh reached Bagicha, where it received orders to speed up the advancement and reach Skardu by 29 April. He replied promising to do his best to carry out the order, but he remained halted on 27 April due to lack of transport. Coutts also remained immobile at Parkutta, due to hostile opposition and the absence of coolies, although he had received orders from Lieutenant Colonel S.B. Singh to push on one stage further. On 28 April, Singh marched up from Bagicha to near Parkutta, where Coutts was greatly relieved to hand over control of the operations to him.

Meanwhile Lieutenant Colonel Kripal Singh was slowly wending his way through the gorges towards Dras and Kargil. He was in an unhappy situation, as HQ 163 Brigade continued to send orders direct to Major Coutts and Lieutenant Colonel S.B. Singh without even informing him, who was supposed to be the commander of the entire relief column. Lieutenant Colonel S.B. Singh and Major Coutts he claimed, ignored him and kept him in the dark, maintaining no communication with him at all. He felt that, although on his strong representations, Brigadier Lakhinder Singh (Commander 163 Brigade) and Major General Kulwant Singh had made him the column commander in place of Lieutenant Col. S.B. Singh, he had no effective control over the column strung out over 160 km of track. On 27 April, in order to concentrate his force and acquire effective control, Lieutenant Colonel Kripal Singh ordered that no troops should be moved without his prior sanction. On 1 May, Lieutenant Colonel S.B. Singh reported these developments to 163 Brigade and asked for a clarification of the position. In reply, 163 Brigade informed all concerned that Lieutenant Colonel S.B. Singh would be in independent command of the foremost column, and that Lieutenant Colonel Kripal Singh would take over from him as soon as he (Kripal Singh) caught up with the advance guard. Kripal Singh had reached Kargil on 30 April, and he was ordered to hurry up and to push forward at the best pace possible. 163 Brigade also impressed on everybody the need to work in co-operation. But it produced no improvement, and friction continued between the two senior officers. The raiders were all around the isolated parts of the column, no coolies or pack ponies were available, and ration stocks were getting low at Parkutta. The advance was at a standstill. On 4 May, 163 Brigade finally ordered that Lieutenant Colonel Kripal Singh alone would be in command of the entire relief column and responsible for saving the starving garrison of Skardu; Major Coutts would remain with him merely as an Indian Army observer and Lieutenant Colonel S.B. Singh would return to Srinagar. About this time (1-2 May) the troops of the column were located as follows :

Four platoons of 5 J & K Infantry, plus one detachment of 3-inch mortars, plus some troops of 7 J & K Infantry	-	Parkutta
Tac HQ 5 J&K Infantry, Signal platoon, one company and one section MG	-	Kargil
One platoon plus a detachment of 3-inch mortars	-	Gumri

Six platoons, a 3-inch mortar
and a section MMG Sonamarg

The forward troops at Parkutta were deployed on both the banks of the Indus. Four platoons were on the north bank of the river, holding platoon picquets, with a reserve platoon at hand. On the South Bank of the river, there were four picquets and the reserve was located nearby, and one platoon had accompanied Major Coutts to Kargil to fetch rations.

By then, these operations were reaching a crisis. In Skardu Lieutenant Colonel Thapa's position was deteriorating day by day. On 15 April, he had reported that all his 3-inch mortar bombs were expended; 2-inch mortar ammunition was also nearing completion and even rifle cartridges were in short supply. He complained, quite justifiably, that the progress of the relieving column was very slow, and feared that all his ammunition might be exhausted before he was reinforced. But more and more troops were coming up to Kargil and it seemed clear that, soon their preponderance in numbers and fire-power would be so great that Skardu would be relieved inspite of all the difficulties and handicaps.

The hostile leaders also recognised their danger. If they went on waiting to be attacked, they were lost. But, they had captured the initiative and moral ascendancy during the skirmishes, and had obviously a poor opinion of their opponents in the Kargil sector. So, their commander decided on a brilliant and daring stroke, designed to destroy the Indian forces before him, capture Kargil itself and make the fall of Skardu virtually certain.

This imaginative plan was based essentially on the fact that the Indian line of communication from Srinagar to Skardu ran parallel to the front for much of the distance. In the Skardu sector the enemy commander had about 800 men at his disposal. Of these, about 200 men were to be left to besiege Skardu and to prevent the rapid advance of the relieving column beyond Parkutta. The rest were to be used for a right hook against the Indian line of communication from Sonamarg to Kargil. The base for this right hook was Chilam Chowki on the main Gilgit - Srinagar route. The preliminary operations were well conceived and planned in detail. From Chilam Chowki, a force of 250 men was to march South; cross the Burzil Pass; occupy Gurais; cross the Razdinangan pass and demonstrate against Bandipur. This would distract the attention of the Jammu and Kashmir forces, and might even result in Indian reserves being rushed off to the quarter. Then, the raiders were to hurry back to Gurais, and set off towards east for the real stroke, leaving about 150 men near Gurais to protect the Burzil route and the base of operation.

The operation was to be carried out by four separate columns with distinct, but coordinated roles. The first column of two platoons was to advance from Gurais via the Tilel Valley and Baodab, then turn south and strike the Indian line of communication near Gund. Its task was to destroy the wooden bridges on the route, thus preventing the rapid movement of troops from Srinagar to Dras and Kargil. The second column also of two platoons, was to accompany the first column upto Baodab; and then, keep going east; cross the main Himalayan range by the Kaobal Gali; move down the Muski Nala and cut the line of communication near Pindras. The third and the fourth columns were to set out together from Chilam Chowki, climb up to the Deosai Plains and go down the Shingo valley to Gultari. From there, the third column of about 100 men, was to march to Dras and capture it. The last column, of about 150 men with two 3-inch mortars, was to advance down the Shingo river and capture Kargil itself. All the columns were to strike at their objectives on the same day - 10 May, 1948, and their marching tables were arranged accordingly. Meanwhile, the raiders near Parkutta were to fall back gradually towards Skardu, drawing the main Indian column after them and away from Kargil.

In conformity with this plan, the faint attack against Bandipur was carried out on 28 April. The main force of 350 men concentrated at Chilam Chowki on 29 April. The plan was re-examined and finalised. Surprise was to be their decisive advantage, and the difficult route and intense cold were their major hurdles. Each man of the attacking force was given one week's ration and 150 rounds of rifle ammunition to carry, apart from his blanket. Coolies and sledges were used to carry the reserve stock of 10 days ration, spare ammunition, and the two 3-inch mortars. The main force moved out of Chilam Chowki at 0100 hours on 1 May For three days and nights it advanced through a desolate wilderness of rocks and snow at an altitude of about 4270 metres. No shelter or fire wood was available, and the men subsisted on their pre-cooked rations. A snow blizzard caught them on the second day. .But, they were hardy locals used to these conditions, and on the Deosai Plains the going was quite easy. On the evening of the third day, they reached Gultari having lost 4 men dead in the snow and with about 60 other casualties due to frost-bite. But, Gultari was comparatively sheltered, fuel was available, and beyond it the route was easier.

The first and second columns left Gurais on 3 May. In the upper Tilel valley the snow was deeper because of the heavy snowfall. They found the route exceptionally difficult; and the second column, aimed at Pindras, was delayed by a day in its march. The other three columns harboured in the night of 9/10 May within easy striking distance of their objectives.

With the coming of dawn on 10 May, the first column set fire to several bridges between Sonamarg and Kangan. The second column was then struggling in the snow about

48 km from Pindras. The third column launched an attack against Dras. But the garrison, although taken completely by surprise, promptly manned its defences and beat off the raiders.

The main attack at Kargil was completely successful. First, the raiders crept up to a section of Gorkha troops guarding the Kharal bridge across the Shingo river, less than 16 kms from Kargil. Not one man out of this platoon was allowed to escape and warn the garrison at Kargil. Then the raiders attacked Kargil. The garrison, outnumbered and taken by surprise, could offer no effective opposition. The raiders had sneaked around and occupied positions on the hill behind the town, and their 3-inch mortars played havoc with the defenders running to their positions. The fight was soon over, the garrison scattered and escaped by batches, some along the Indus towards Leh: others up the Suru to Pahalgam.

The last act of this dismal tragedy was soon played out. On 4 May, '63 Brigade had informed Lieutenant Colonel Kripal Singh that, according to a reliable report, the raiders were advancing from Gurais area to Dras, and had asked him to establish a firm base at Dras with a garrison of one company. Accordingly, one company (less one platoon) of 5 Kashmir Infantry which had then reached Dras under Captain Kashmir Singh, was ordered to halt there and form a firm base. But, with the fall of Kargil, the main column between Parkutta and Bagicha was caught between the enemy in front and the enemy behind. Under energetic leadership, it could still have turned the tables by attacking and defeating the two hostile forces in turn. But, the situation was confused and no detailed account of what happened appears available. The tail of the column was near Bagicha under Lieutenant Colonel Sampuran Bachan Singh. Headquarter 5 Kashmir Infantry and about two platoons with a section of MMGs and a detachment of 3 inch mortars were at Minigund near Tolti, about 22 km from Bagicha, under Lieutenant Colonel Kripal Singh. The head of the column was near Parkutta, about 25 kms further on, where there were over three companies of troops and two detachments of 3-inch mortars. The raiders had infiltrated in between and opposed every movement of any part of the column. Thus split up, immobilised and dispirited, the column fell an easy prey to the raiders, although in theory it was still very strong and capable of overcoming all opposition.

However, Lieutenant Colonel Sampuran Bachan Singh and Major Coutts fell back hurriedly from Bagicha soon after the fall of Kargil. Then, their troops were attacked and quickly dispersed, the two officers saving their lives by swimming across the ice-cold Indus and fleeing up the valley. They reached Khalaste bridge on the Leh route on 13 May.

After the exit of Lieutenant Colonel Sampuran Bachan Singh and his contingent there were still about 600 men of 5 and 7 Kashmir Infantry left under Lieutenant Colonel Kripal Singh between Tolti and Parkutta. On 12 May, Lieutenant Colonel Kripal Singh's message read - 'Completely surrounded. Enemy attacking all sides. Cut off from all sides. Situation grave.' After thus repeating himself several times, he went on to ask for immediate air strikes to extricate his force and for supply drops from the air. Major-General Thimmayya passed on his request to Western Command adding that, unless the column was given immediate support from the air or permitted to withdraw, it was quite likely to surrender. On 15 May, Western Command sanctioned a withdrawal with a view to the recapture of Kargil, and General Thimayya ordered all the troops in Skardu and Parkutta and Tolti to fight their way out and concentrate at Olthingthang near Marol. As related below, the Skardu garrison was exempted from his order on Thapa's representation, but Lieutenant Colonel

Kripal Singh started pulling back as ordered. He had supplies weighing about 18750 kg, and no local porters were available. The raiders were harassing his column from either bank of the Indus. Still, Kripal Singh sent orders to his isolated parties of troops. They retreated from Parkutta on 20 May, and left Tolti on the night of 21/22 May. By 1200 hours on 22 May, the column was near Kharmang bridge. Suddenly, the raiders attacked it from all sides - the column had walked into an ambush again. The local coolies immediately threw down their loads and fled. The column suffered about 200 casualties, lost its baggage and heavy equipment, and practically ceased to be an operational force.

But it still retained some cohesion, and, since Kargil had already fallen, it struck out south-west into the hills. After terrible privations over the trackless snow and avoiding the routes, these men reached Faranshat in the valley of the Shingo river. Misfortune again overtook them there. The men were crossing the turbulent river by an old rope bridge when the enemy suddenly appeared. In the confusion, the rope bridge gave way and the survivors split into two parties, fleeing westwards along either bank of the Shingo. The party on the right or south bank of the river, consisting of A and B companies of 7 Kashmir Infantry found going very tough. It was decided, therefore, to turn due south and make for Dras. But on 1 June 1948 it was again encountered by the raiders when only a few kilometres from Dras. The men were starving and completely done up by the terrible ordeal of the previous fortnight. They were practically unarmed and were easily overpowered, and only 2 survivors reached Srinagar on 8 June to tell the tragic story.

The party on the left bank of the Shingo river consisted of about 150 men and was led by Lieutenant Colonel Kripal Singh himself. It shook off the pursuit, crossed the Shingo near Gultari and again struck into the mountains to the south. Lieutenant Colonel Kripal Singh and his 150 exhausted men reached Sonamarg safely in June. Thus ended the strong column of which so much had been hoped.

Dras Falls to the Raiders

Efforts were then made to relieve Dras. Indian troops were then located as follows :

Two Platoons of 5 Kashmir Infantry at Dras ;

One Company of 5 Kashmir Infantry at Gumri ;

One Company of 5 Kashmir Infantry at Baltal ; and

One Company of 5 Kashmir Infantry at Sonamarg.

These troops were obviously insufficient for clearing the vital track upto Dras. So, leave alone relieving Skardu, even Dras was cut off. Many efforts were made by 1 Patiala to link up with Dras. Eventually, even Patiala was isolated by enemy who reached and dominated the road leading to Zojila. At this stage 1 Patiala was ordered to fall back and hold Zojila.

The above account from the official histroy of the Indian Army clearly shows how poor leadership and lack of tidy command and control structure can be disastrous. Lieut. Colonel Sher Jung Thapa continues with his story:

With the withdrawal of the Indian troops to Zojila, the relief of Skardu became quite out of question for sometime. The enemy strike against Kargil and Dras had proved a brilliant success, and they confidently expected Skardu to fall into their hands like a ripe fruit.

Slowly we were left with an area of not more than 1500 yards by about 500 yards under our control at Skardu with enemy positions only about 50 yards or less from each picquet. The enemy changed his tactics from attacks to that of infiltration into our headquarters, through gaps. This was met by filling up of major gaps and constant night patrolling. By this time our supply and ammunition and manpower situation had become critical. Hundred shells from them were replied by us only with a shell or two.

The ration position was even more grave, almost bordering on starvation. The smokers had to resort to smoking tree leaves and rice husk. Dogra soldiers are almost addicted to smoking. The more enterprising ones in the garrison rolled up tea leaves in mulberry leaves to make cigarettes. Quite a few got blisters on their lips. One day I found my Dogra orderly having a bandage on his bicep. I was curious as to how he had got wounded in the relatively safe position of the headquarter. When I persisted in my questioning the truth came out. He and some of his cronies who were desperate for a good smoke had sneaked out from the defences to a nearby cluster of civilian houses, for a good smoke from a hukka (hubble-bubble). While they were sneaking back, an alert sentry took a pot-shot in the direction of the shadowy moving target and my batman was the one who caught the bullet on the fleshy part of his bicep. It was after this incident that I made a special request for air drop of some cigarettes.

The effect of virtual starvation had a serious adverse impact on the morale of troops. Enemy activity on the line of communication behind us had considerably increased . The enemy realising our weak condition, contented itself with occasional jitter or nibbling attacks with a view to exhaust our ammunition.

On 16 May I received a message from Sri Division, asking me to withdraw to Olthing-Thang (Near Kargil) with refugees and as many arms and ammunition as could be carried, with the intention of recapturing Kargil which had already fallen. The same message was addressed to Biscuit Column for similar action. In reply I requested Sri Division Commander, General Thimmayya to re-consider his order in view of the following difficulties :

a) The refugees who included children, old men and women would not be able to endure the hardships to be encountered on the way back. It would be inhumanly cruel to leave them at the mercy of a fanatic enemy.

b) The enemy was already in occupation of strong positions on the line of withdrawal right upto Olthing-Thang which was about 80 miles. It was not possible to break through those positions, without suffering very heavy casualties.

c) We had about a dozen stretcher cases which could not be carried for want of transport, nor could they be left behind.

d) We had not been left with any ammunition for supporting weapons.

I also talked to General Thimmayya personally on the radio set and he conceded to my request and later cancelled the withdrawal orders. He promised me supply of rations and ammunition by means of air drops besides air strikes on enemy position. The air raids started as promised. However, the supply dropping area was too narrow, as a result of which 50 per cent of the containers fell into enemy hands.

In compliance with orders of withdrawal from Sri Division the Biscuit Column withdrew and remained out of touch for many days. The first information about this column came to me when Lieutenant Colonel Kripal Singh and remnants of the column reached Srinagar. This news completely shattered the morale of my men who were left without any further hope of receiving men and materials from Srinagar.

On 17 June we saw Sepoy Amar Nath of 5 Kashmir Infantry captured by the enemy from Parkutta. He came towards our defences carrying a white flag. He was sent by the enemy to persuade us to surrender. He carried a letter addressed to me wherein the enemy had promised good treatment in the event of surrender. Besides this, the man was properly tutored and indoctrinated to demoralise our men. This letter signed by Colonel Shahzada Mata-Ul-Mulk, brother of Mehtar of Chitral, who had been an officer in Subhash Chandra Bose's Indian National Army during World War II. The letter is reproduced below. Needless to say, that I rejected this overture by the enemy:

To:

The Officers and Men of Kashmir State Forces, Skardu Grn.

From :

Colonel Shahzada Mata-Ul-Mulk, Comd. Azad
Chitral Forces, Skardu (.)

One (.) All Attempts to relieve your Grn by Brig Faqir Singh, Lt Cols Kripal Singh and Sampuran Bachan Singh have resulted in absolute failure in numerous killed and prisoners taken (.) Azad Forces are now operating in Kangan, Sonamarg and Bandipur area also in some cases within 15 miles of Srinagar (.) Two (.) You have done your duty as every soldier should do (.) Now that it is clear that no relief can reach you in this mountainous area there is no doubt about the ultimate fate of your garrison and I am quite determined about it (.) It is no use to carry on a struggle which will result in your total annihilation (.) Three (.) I therefore advise you to lay down arms and I take full responsibility to give protection to one and all (.) You must believe me and trust me as I am not only a soldier but also possess royal blood (.) I have given instructions to my officers and men that any one approaching with a white flag will not be fired at but taken into safe custody (.) Four (.) Lastly as a proof of my goodwill I wish to inform you that not a single Sikh or Hindu resident of Chitral has been hurt and not a single non-Muslim property looted or damaged and uptil now they carry on their business as if nothing at all had happened (.) I therefore advise you again to lay down arms and thus save your lives (.) An officer

The mounted Wing of National Guard raised out of volunteers at Leh to resist Pakistani invasion in 1948. Subedar Bhim Chand, Vir Chakra in Command.

should accompany back the white flag if you consider my words sincere and honest (.)

The morale of our men shot up again on 19 June when two of our aircraft (Tempests) appeared and strafed Point 8853 and inflicted casualties on the enemy. Air dropping was not much of a success because Tempests could carry only limited quantity of supply, ammunition and ration, 50 per cent of which fell into enemy hands. Besides, the frequency of drops was low. It was impossible for the Tempests to maintain the garrison, our minimum daily ration requirement being 800 lbs.

The enemy was increasing pressure again on Skardu defences. Attacks began to come with greater frequency and intensity, but they were always successfully repulsed. Enemy had no doubts about the fighting qualities of our men though they were fighting under such a great disadvantage.

On 9 August 1948 the enemy brought up two 3.7 Howitzers and shelled our positions from 17-00 hours to 18-00 hours. They must have fired about 300 shells. We could only retaliate with LMG and 3-inch Mortar, but that did the job and silenced the gun.

Again, in the evening on 12 August, enemy opened heavy artillery fire followed by an attack with two companies. The enemy was able to force an entry in the picquet area, face to face fighting took place and hand grenades were freely used by both the sides. After about three hours of battle the resolute and brave boys defending the Skardu stronghold beat them back. The enemy left behind a few dead and wounded. Our casualties were less as compared to the enemy, but this battle was very expensive for us. We practically used up all our ammunition and our energies were completely exhausted.

Internal communication with our picquets was by means of runners. Normally, it was carried out during darkness because of the enemy's commanding position at Point 8853. I was connected with the Divisional Headquarters with the radio set which I had kept with such care. I was in touch with the Division throughout, except for one occasion for a couple of days, when it affected the morale of a few refugees to some extent.

Long siege of the garrison, illness from malaria, dysentry etc. combined with shortage of food was telling upon the health of Jawans, mentally and physically. The number of sick persons including refugees was daily increasing. Enemy was intensifying attack, resulting in increasing number of casualties. Many of the wounded died because of non-availability of medicines and surgical instruments.

My doctors did what best they could to save them. After a continuous barrage of SOS requests for dropping medicines, an air-drop took place. What fell within our defences had only 12 phials of penicillin, that too were date expired. What a pity !

SKARDU FALLS - 14 AUG. 1948

On night 13/14 August the enemy launched another determined attack outnumbering us 5 to 1. In this battle we used our last box of reserve ammunition. Enemy was able to run over a portion of our position. Everyone knew our plight, and there was panic and chaos all over. The women started committing suicide by jumping into river Indus and poisoning themselves in order to save their honour. There was an instance where a girl jumped thrice in Indus river to kill herself, but everytime the waves carried her back to shore. Similar panic spread among the troops too. Some of them left the position and deserted, later they were killed by the enemy except a lucky few who reached Srinagar.

My radio operators had deserted as enemy was approaching. I wanted to contact Division Headquarters to report full details of the situation. I was lucky. My attempt bore fruit and at about 05-00 hours the Divisional set came on the air. My speech being weak, the operator at the other end rebuked me, taking me also to be an operator. He wanted me to tune the set properly. But I was afraid that even a slight fiddling might put me out of net. With God's blessing the operator at Leh came to my rescue as we were on the same net. I asked him to fetch Lieut. Colonel Prithi Chand on the set. He came and I explained the whole situation to him. He asked the operator at the Divisional Headquarters to tune his set to mine which he did, and thus I was able to speak to the General Staff Officer Lieut. Colonel Sri Ram Oberoi MC, because the GOC was not available. I explained the situation to him and told him that I had to hold Skardu to the last man and last round, as ordered by General Thimmayya. My troops had gallantly fought under very adverse and totally isolated conditions and held Skardu for 6 months and 3 days. Now I was left with no other alternative than to surrender. After a while he confirmed that I may surrender - he had obviously consulted General Thimmayya on the telephone. I then said 'Good bye - Good luck' and perhaps also conveyed my disappointment about the non-fulfilment of the promise to relieve us. This was the end of the gallant siege of Skardu. The surrender was followed by mass murder and rape. All the Sikhs were shot dead. Captain Ganga Singh, my Adjutant, was tied, laid on the ground and shot. The only Sikh who escaped was Kalyan Singh who was my Orderly and was staying with me.

When we as prisoners of war reached Attock fort, where the prisoners of Jammu and Kashmir were held, General Gracy who was then the Army Chief of Pakistan came to visit the fort. He inquired from the fort commander about me and came straight to my room. After meeting me he went to inspect the bath room and opened the tap. There was no water. He was angry with the Pakistani Camp Commandant for the shabby and bare furnishing of my room and immediately ordered a proper bed, writing table and other necessities. It was very nice to meet him after a lapse of almost 25 years. But I wondered how did he come to know that I was at Attock.

I was told the truth many years later when I met General Kulwant Singh in 1955, while I was commanding 120 Infantry Brigade. When we surrendered at Skardu, a message was intercepted by our signallers. It had originated from the Commander of Azad Chitral Forces (probably Shahzada Mata-Ul-Mulk who had sent me a message on 17 June to induce me to surrender). He was asking higher headquarters permission to kill Lt. Col. Thapa and all the prisoners of war at Skardu who had inflicted many casualties on them. This intercepted radio message was promptly sent to Army Headquarters at Delhi. On seeing it General Roy Bucher, the British Indian Army Chief in India immediately rang up his counterpart General Douglas Gracy in Pakistan. Gracy acted with despatch and issued orders to stop any further killing of prisoners of war. My old friend had saved the life of all of us and thus I came to know how Gracy was aware that I was at Attock. He had obviously kept a tag on our movement. After General Gracy's visit, the attitude of Pakistani captors completely changed and we were treated very well. Let us now move to Colonel Prithi Chand's role.

CHAPTER 5

Colonel Prithi Chand, MVC, At Leh

We read in Chapter 1 how in 1937 Colonel Prithi Chand promised his family Guru to protect the monastery at Hemis, near Leh, when in danger. Also how, by 1947, he had forgotten all about it, and his late Guru had to remind him, in a dream, to fulfil his promise.

Major Prithi Chand was in the second Battalion of Dogra Regiment being commanded by Lt. Col. G.G. Bewoor (later General and the Indian Army Chief of Staff). This Battalion was a part of 168 Brigade pushing the Pakistani invaders out of Kashmir valley and, by the end of November 1947, had reached Baramula. It was here that Prithi Chand realised that Ladakh could fall to Pakistani invaders any day.

With the closure of Banihal pass during the winter, the build up of Indian Forces in Kashmir valley was only by air. Slowly Headquarters Z Brigade was introduced into the valley to reinforce 168 Brigade. All the forces in Jammu as well as in Kashmir were being controlled by Headquarters

J&K Force under Lt. Gen. Kulwant Singh located at Jammu. On reorganisation of the J&K Force, Srinagar Division and Jammu Division were raised to conduct operations in the two geographic regions of the State. Major General (later General) K.S. Thimmayya reached Srinagar on 4 May 1948 to take over the command of Srinagar Division.

In second Dogra, there was a rifle company of Lahauli and Ladakhi soldiers. At the end of the Second World War, most of them proceeded on release. Apart from Major Prithi Chand, Major Thakur Khushal Chand, his cousin Subedar Bhim Chand and 15 other ranks, were the only Lahaulis and Ladakhis who remained in the Battalion. Prithi Chand gathered all of them and discussed the situation in Ladakh. They agreed that, if early action was not initiated to despatch troops to Ladakh, it would be attacked and plundered by the invaders. So, they decided to request the Commanding Officer to send them to Ladakh to reinforce only a platoon of about 30 men of State Forces stationed there. During December, Prithi Chand vividly described the situation in Ladakh to Colonel Bewoor, and conveyed his concern for the people there and requested troops to be sent there; only one platoon of 6 J&K Infantry at Leh was utterly inadequate to defend Ladakh. The CO of this Battalion, Lt. Col. Sher Jung Thapa had already moved to Skardu. There was also approximately a platoon strength under Captain Bhagat Singh at Kargil on the line of communication between Srinagar and Ladakh.

The Zojila pass was closed due to heavy snowfall in early January 1948. Some State Force troops which were despatched from Srinagar, were already stranded at Sonamarg due to heavy snowfall. Colonel Bewoor consequently said that he could not spare them to go to Ladakh because he needed them at Baramula. Also, that there were no orders from higher authorities to this effect.

In the meanwhile the Young Buddhist Association of Ladakh had organised themselves to defend Ladakh by calling for volunteers. They sent a telegram to Prime Minister Nehru: "We have arrived with bows and arrows to defend Ladakh. Request, send arms and ammunition and reinforcements." The Prime Minister realising the importance of Ladakh, which had borders with Tibet, China and Russia and the Buddhist population and their monasteries exposed to enemy threat, ordered the Defence Minister to arrange for the defence of Ladakh. The Chief of the Army Staff was asked to take immediate action. He ordered General Kulwant Singh, Commander of Forces in Jammu and Kashmir to handle the situation. General Kulwant in turn instructed Brigadier L.P. Sen, the then Commander of 161 Infantry Brigade in Kashmir with its Headquarters at Baramula, to be responsible for the defence of Ladakh. These instructions were issued in January 1948.

Before partition, Brigadier Sen and Lt. Col. Bewoor belonged to the Baluch Regiment. They often used to meet and discuss operational matters together. Brigadier Sen discussed the issue of Ladakh with Bewoor. Since he had only one Brigade to operate in the entire Kashmir valley, it was not possible for him to undertake this assignment. He was still short of troops. The defence of Ladakh seemed to be impossible because the Zojila pass was closed and the entire valley had heavy snowfall. So, the responsibility to defend Ladakh was assigned to the Jammu and Kashmir State Force under Z Brigade (later designated as 163 Brigade). We have seen in Chapter 4, how the main focus of attempts of this Brigade was to reinforce Skardu.

Colonel Bewoor nevertheless informed Brigadier Sen that Major Prithi Chand of his Battalion, Major Khushal Chand, a JCO and 15 other ranks of Lahaul and Ladakh, who were all Buddhists, had requested to be sent to Ladakh, for defending their families and the monasteries.

The Brigade Commander was reluctant to despatch a handful of men to such a distant place without proper support. However, considering the importance of the mission, he agreed to permit the contingent to go to Ladakh, provided all members of the contingent were volunteers. All of them unequivocally confirmed that they were happy to go in this mission. The Brigade Commander finally gave his consent and Prithi Chand was directed to work out requirement of arms, equipment, clothing, ration, etc., needed. The group then moved to Srinagar to organise themselves for the expedition.

This group was named X Column. The following personnel constituted the party :-

1.	Major Prithi Chand, Column Commander	)
2.	Major Khushal Chand	)
3.	Subedar Bhim Chand	)
4.	Naik Mindol	)
5.	Lance Naik Nawang	) From Lahaul
6.	Lance Naik Karam Singh	)
7.	Sepoy Punchog	)
8.	Sepoy Sonam Angrup	)
9.	Sepoy Tobge Ram	)
10.	Sepoy Sonam	)
11.	Sepoy Lobzang	)
12.	Sepoy Tashi Chhering	) From Ladakh
13.	Sepoy Tashi Zanskar	)
14.	Sepoy Bahadur Singh	) From Kinnaur
15.	Sepoy Dalip Singh	) From Kangra
16.	Sepoy Patwari Mal	From Kangra (Signal Regt.)
17.	Shri Sonam Norbu, Engineer J&K	

The party was equipped with one radio set No.22 and each individual was issued one sheep skin jacket, a cap and a pair of sheep skin gloves. They had one month's ration, some medicines, vitamin tablets and 50 extra rifles and ammunition, in addition to their personal arms and ammunition.

Brigadier Lakhinder Singh of Z Brigade issued the following orders to Column X :-

a) You will move to Leh.

b) You will raise an irregular force and train them.

c) You will defend Ladakh by stopping the advance of the enemy and protect vital grounds and bridges.

d) You will construct an airfield with the help of Mr. Sonam Norbu, Divisional Engineer.

e) You will keep the enemy away by using guerrilla tactics, till such time the regular troops arrive there.

f) You will keep the lines of communication open to Kargil.

g) You will live on local resources, through local purchase.

It is best if, from here onwards, we read the story in Prithi Chand's own words:

With the above orders we prepared for our move and left Srinagar on 16 February 1948 in three ton vehicles and reached Gund in the evening. Due to heavy snowfall the vehicles could move only upto about two miles short of Gund. We, therefore, had to carry our baggage and complete the journey on foot.

At Gund we had heavy snowfall for four continuous days. I asked the Labour Officer, Mr. Ghulam Mohammad, to provide the column with porters. He, however, was not co-operative. As a result I rang up Shri D.P. Dhar of the

National Conference who was one of the many stalwarts assisting the Indian Army as Liaison Officers at Srinagar and complained to him about the attitude of Mr. Ghulam Mohammad. He immediately spoke to Ghulam Mohammad - who then provided the, column with the needed porters. On 24 February we left for Sonamarg and halted there. There Brigadier Faquir Singh of J&K Forces with one rifle company earmarked for Skardu was held up for nearly a month. When I told him of our plan to leave the next day for Baltal and move to Leh, he said that it was impossible to cross Zojila pass for at least a month. He said there was one Mr Habib Lone, an experienced mountaineering guide who had considerable experience of snow. It was his advice that Zojila pass would be fit to cross only after a month. I sent for Mr Lone and enquired from him about his experience in snow and passes. The conversation, went like this :-

Lone : I have participated in several mountaineering expeditions and climbed many peaks including Nanga Parbat.

Prithi Chand : Did you attempt to climb these mountains in summer or in winter?

Lone : In summer only.

Prithi Chand : Lone, I have crossed mountains and passes several times during winter. On many occasionsI have crossed the Rohtang pass linking Lahaul and Manali during the months of February and March. I am sure you know that Rohtang pass at 13,400 is more than two thousand feet higher than the Zojila.

I then requested Brigadier Faquir Singh to send Lone with us upto Zojila, and that we would leave for Baltal which is at the foot of the final climb to Zojila, the next day.

The next day Column X left for Baltal while it was still snowing heavily. We had great difficulty in locating the Rest House at Baltal which was fully buried in snow, except a small portion of the chimney which was sticking out above the snow. We dug the snow and entered the rest house. The porters seemed to be extremely frightened of the snow. Mr Habib Lone informed me that the porters were planning to desert at night. To ensure that this did not happen, I established a night guard at the exit to stop the porters from running away.

Next day, on 26 February, strong winds and blizzard continued. Visibility was hardly 10 yards. However, we left Baltal. While climbing up to the pass, we encountered strong wind and sleet which bruised our cheeks and noses. Many a time, some members of the column suggested to me to stop the climb and return to the camp.

Giving up so easily was unthinkable. So I prayed to God for help. There were thousands of people in Ladakh who were restless and scared of the invaders. It was important for us to reach Ladakh before the Pakistani invaders. If we delayed our advance and reached late, after the Pakistanis, we would repent for the rest of our lives. I prayed from the bottom of my heart. I offered my life to the Almighty, but prayed for the safety of the innocent population of Ladakh. As I prayed, there was what appeared to be a miracle. Immediately the clouds scattered, the winds stopped and bright sun started to shine over us. With cries of "Allah Hu Akbar", "Bhagwan ki Jai" and "Knochhog", we moved on, crossed the pass and reached Gumri at the far end of Zojila pass. Here, the tough porters from Matayan and Dras had come to meet us and they carried our baggage and the Kashmiri porters returned to Gund.

When we were half way between Gumri and Machoi, a huge avalanche came roaring down the hill. Most of us were buried in waist deep snow. We could not even run away. But providentially the fury of the avalanche abated, and we slowly extricated ourselves. We reached Machoi and spent the night there in the few huts that were there. At Machoi there was a Kashmiri Pandit who was the single line telephone operator between Ladakh and Srinagar. He advised us not to move too early in the morning and start only after it was bright sunshine. Before that the wind in this area was too strong and there was always a danger of avalanches. Thanks to his advice, the column advanced only at about 10 am, when the strong wind and blizzard had stopped. On the way, we did see signs of fresh avalanches before reaching Matayan, the first village of district Ladakh. I sent a message to Srinagar about our crossing the pass. Back came a congratulatory reply, "Hats off to you and wish you all success. Heartiest congratulations".

Next day, the column left for Dras and rested for a day there. The Head Master of the school told me that the temperature at Dras had touched minus 60 degrees celsius, it was well in keeping with its reputation as the second coldest inhabited place in the world.

On 1st March the column reached Kargil and met Captain Bhagat Singh of Lieut. Colonel Sher Jung Thapa's Battalion. He informed us of the situation at Skardu. The enemy had surrounded Skardu and was repeatedly attacking it. The invaders were expected to reach Kargil before long. So I decided to leave for Leh, after only a day's rest.

On 3 March the column left for Mulbek where we reached in the evening. This is the first Buddhist village on this route. When the villagers came to know that Indian

Army troops were arriving, the entire village came out to welcome us. They had Joss-sticks, curds, milk pots and Arak in their hands. Some of them even prostrated on the ground. They had no hope of surviving, but when they saw the Indian Army, they felt relieved and happy. One old Lama told me that as he was lame and could not run away, his Muslim neighbour had taught him the *Kalma* and said that he would introduce him to the invaders as his elder brother. Such was the culture of love and tolerance among people of Ladakh.

The next day the column halted at Lamayuru near the famous monastery. Here too, we were given a warm reception by the villagers and the Lamas. After another three days march through the villages, where similar revival of morale took place, we reached near Spituk Monastery, where a few young Buddhists entertained us to tea before we could proceed to Leh. The column was received near Leh by about a hundred men on horses. They were mostly representatives of the Chinese traders, Indian traders, representatives of Hindus, Muslims, Christians and Buddhists and civilian officers. The members of the column were lost in this crowd. At the entrance to Leh Bazar, a local band was playing welcome tunes. The column was invited to tea at the residence of Mr. Donyod Sahib, father-in-law of Mr. Sonam Norbu. The senior Tehsildar of Leh suggested that we should stay in the old mud fort of Leh, but I refused and decided to stay in the residence of the British Joint Commissioner. In front of this house there was a Union Jack hoisted on a wooden flagstaff.

Next day the two Tehsildars of Ladakh, both Muslims, called on me. Kalon Rigzin, head of Buddhist Association, Khawaja Abdullah Shah, the Raja of Stok, head priest of Leh Monastery, Shri Sridhar Kaul ADI of Schools, Indian, Tibetan and Chinese traders also visited me. I thanked them for the cordial and warm reception they had accorded to the troops and told them that on 13 March the Indian Tricolour would be hoisted on the flagstaff. I would then inform them of Indian Army's mission and future plans. I also requested them that the maximum people of the town should attend the ceremony, the Lamas may be requested to pray on that day, and the local band should be in attendance. The Indian traders asked for permission to organise a tea party for all those who attended the ceremony which was granted.

On 13 March, at 11 a.m., there was a huge crowd in front of the Headquarters to witness the flag hoisting ceremony. One platoon of State Forces and men of X Column stood in front of the flag. The Lamas prayed, and while they were playing puja cymbals and drums, the soldiers presented arms and the Union Jack was respectfully lowered. Following this the National Flag was hoisted with soldiers presenting arms again. I then addressed the crowd and said, "we have been sent here to help you to defend Ladakh. We will give you training and then organise our defence. As you know, all the passes are closed, it is difficult to get reinforcements to this place. Till such time the Army arrives, we have to defend Ladakh at all costs. I therefore, request you to give full co-operation to us in recruiting and training your young men. Without your co-operation we will be unable to carry out our mission. Mr. Sonam Norbu is here with us to construct an air field. As soon as it is ready we will get our supplies, arms and reinforcements. We request you to help in the construction of this airfield as early as possible. All of you know that in Pakistan thousands of innocent Hindus and Sikhs were killed and their temples burnt. They have also plundered and burnt all Hindu and Sikh houses. In Kashmir valley, the invaders have not spared even the Muslims.

"Now, if you wish to protect the monasteries, save our culture and the honour of our women and religion, then come forward, get training in arms from us and join us to fight the invaders. So far the invaders have not come near us, but who knows what happens tomorrow. It is better to die honourably while defending our country, rather than run away. Your ancestors fought many battles against the Huns, Mongols, Tartars and others. They were very brave people. The same blood is running in your veins. With a little training you will get confidence and be able to face the enemy. While I request you to maintain communal harmony, I warn you that whoever sympathises, propagates or helps Pakistanis will not be spared." I ended my address by thanking the locals for the reception and the ceremony.

In consultation with Mr. Kalon Rigzin (the Wazir of Ladakh) and others, we decided to despatch a few soldiers with some rifles to villages where they would give training to volunteers in handling the rifles. Training centers were established at Nurla, Saspol, Nemo, Chushul and Leh. Even students from schools were enrolled. Chewang Rinchen of Nubra was a student of 8th class (being promoted to 9th) when he joined these guerrilla warriors. He later won two Mahavir Chakras, a Sena Medal and became a Colonel.

Mr Sonam Norbu, Divisional Engineer PWD, belonged to Leh proper. He got down to constructing the airfield. While some labourers were employed, many school boys, Lamas from monasteries, locals of Leh and villagers gave a helping hand in this work. With their help and supervision by Mr Norbu a fair weather airfield was constructed in a few weeks.

While Major Khushal Chand, Subedar Bhim Chand and his men were busy in collecting information about the enemy and organising recruitment of the irregular force, the State Force platoon under Jamadar Thakar Singh was busy in imparting training. I went out touring and visited several villages. I told the villagers how brave their ancestors were, and how they fought and won battles against the invaders under the leadership of Rajas of Ladakh. I made them sing the old victory songs. I told them how Pakistan had killed our men, raped our women and destroyed the places of worship belonging to non-Muslims. I injected into them the idea of 'Do or Die' for defending Ladakh. The population of Ladakh responded magnificently.

VISIT TO NUBRA VALLEY : Along Shyok River

Central Ladakh consists of two main valleys, the Nubra and the Indus. Nubra valley lies across the Ladakh range to the North of Leh. There are three main passes - Khardungla pass (18,500 feet), Digarla pass (17,900 feet) and Changla pass (18300 feet) - to cross the Ladakh range. The inhabitants of Nubra are mostly Buddhists. This valley opens towards Khapalu, which is predominantly a Muslim area. The enemy from Skardu could easily advance through Khapalu and enter the Nubra valley. The people of Nubra naturally felt insecure and, therefore, sent a representation to me for sending troops to Nubra. I appreciated their concern and promised to send some troops as early as possible.

I decided to visit Nubra myself and detailed the State Force platoon under Jamadar Thakar Singh to accompany me to Nubra. I planned to move through the Khardungla pass, but the locals advised me that it was not possible to cross it before June or July. I then thought of the Digarla pass, but the locals again dissuaded me for the same reason. At last we left for the Changla pass. We spent the night at village Sakti. On 23 March we left Sakti quite early crossed the pass and camped at Soltak. Due to the altitude, our last batch of soldiers and porters reached the camp at about 9 am. The Headman of Soltak Mr Tashi, helped us and guided us to reach the camp. At night I issued some local alcohol (Arak) to all those who were used to drinks. Mr Kacho Sahib, Tehsildar, a Muslim from a respectable family, with his staff had also accompanied the group on this tour; naturally, he did not drink.

Early in the morning I woke up and went up to the urinal. When I returned and tried to lie down on the bed, I was breathless and my heart was sinking. The cook woke up Mr Norbu, and told him about my difficulty in breathing. Mr Norbu mentioned that it was due to the altitude. He got a hot cup of tea prepared, which I drank and after some time I slept. When I woke up in the morning I felt better.

When we were about to leave, I was horrified to see that Kacho Sahib and a few of his men had swollen faces. I advised Kacho Sahib and his staff to return to Leh because their health condition was bad, but they insisted on accompanying us. They said that swelling was due to the height and lack of oxygen. When we reached Darbuk their swollen faces became normal. So it was, that those of us who had a drink that night did not suffer! Next day we reached Shyok and stayed there.

SHYOK RIVER

Shyok river flows down from the glaciers of the great Karakoram range and joins the Nubra river at the mouth of the Nubra valley, thereafter it flows down and joins river Indus close to Skardu. In the upper valley of Shyok river, once there was a huge flood which damaged vast areas including many place on the banks of the Indus river in Pakistan. Since then no one had crossed the Shyok river to go to the Nubra valley. We were the first to pass through this narrow gorge by crossing the river several times. We had to cross the river at fords which were deep. The ponies were not able to negotiate these fords. Luckily, a caravan of Central Asian Yarkandi traders with their double hump camels were stranded at Shyok due to heavy snowfall in the Karakoram range. We approached them, and asked them if any of them would transport us to Nubra valley on their camels. They were very happy to do so because there was no grass or shrubs at Shyok to feed their camels, while in the Nubra valley these were available.

We took 22 camels and marched through the gorges fording the river with ease several times a day. On the third day we were received by Kalon Lama of Taggar and the Zaildar of Nubra at Agham. We proceeded to Khalsar and spent the night there. Next day we stayed at Deskit; I left the platoon of State Forces there and arranged for supplies and necessities for them through the Zaildar and Kalon Lama. I ordered Jamadar Thakur Singh to organise defences in consultation with the Zaildar and Kalon Lama. I left this platoon in the valley without any wireless or telephone communication as none was available. I was informed that so far there was no news of invaders advancing towards Nubra, but some elements of Muslim population in Khapalu were in favour of Pakistan and hence infiltration could be expected.

RETURN TO LEH

I, alongwith the civil staff, returned to village Digar on 31 March, and on 1 April we left the village at about 4 a.m. to cross the Ladakh range at Digarla pass. We rode Yaks for a few furlongs, but soon due to heavy snowfall we had to give up the Yaks and walked. It was a steep climb. I started counting my paces while climbing. In the beginning I had to halt and rest after every 200 steps, but as we climbed higher we had to rest after every few steps. While approaching the Digarla pass it became almost impossible to take the eleventh step without a halt. Sepoy Phunchog and one Ladakhi Buddhist policeman were the first to reach the top of the pass. We halted there for the Naib Tehsildar and his staff, who were following us but we saw no sign of them. After almost half an hour I saw them climbing up. I was shocked to see how the civil staff ill-treated the villagers. This officer and his staff had tied ropes around their waist and were being pulled up by a team of five villagers, while they were hanging on to the rope. I don't think these men were even paid for the job, and hence probably this party intentionally stayed behind,

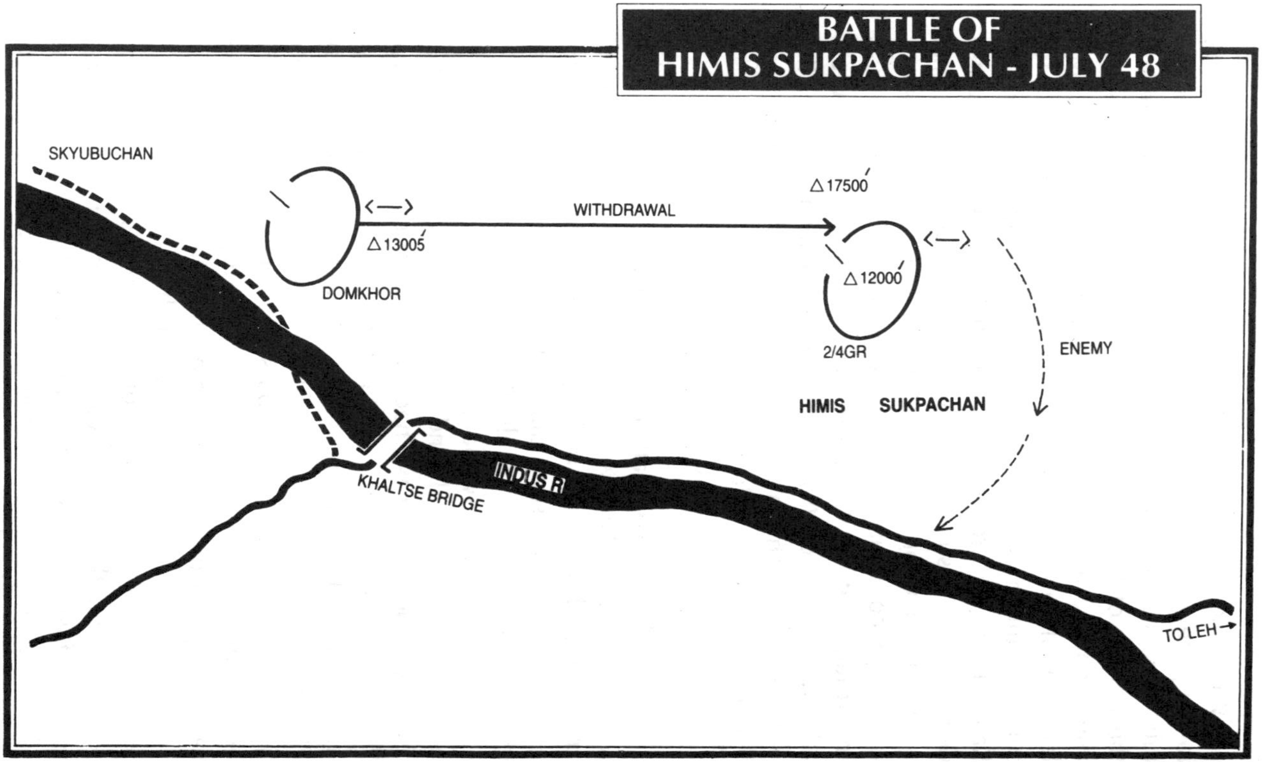

BATTLE OF
HIMIS SUKPACHAN - JULY 48
SKYUBUCHAN
DOMKHOR
△13005′
WITHDRAWAL
△17500′
△12000′
2/4GR
ENEMY
HIMIS
SUKPACHAN
KHALTSE BRIDGE
INDUS R
TO LEH

SKETCH - 4

so that I may not be a witness to it. Anyhow we did not waste time to wait for them. We descended and reached Sabu village at about 8 pm. It was a tough and long journey of about 22 miles. Next day we reached Leh.

INFORMATION AT LEH

Major Khushal Chand informed me that two platoons of J&K State Force had arrived at Kargil and were on their way to Leh. The line of communication from Kargil was so far safe. One rifle company of State Force with arms, ammunition and supplies for Skardu garrison was also on the move to link up with the besieged garrison. They were about 10 miles short of Skardu, when they were ambushed and suffered heavy losses, and thus had to withdraw to Kargil leaving behind their arms, ammunition and supplies. Local porters employed by them had deserted and joined the invaders.

The senior Tehsildar of Leh who belonged to Skardu and the Naib Tehsildar of Kargil who was his brother-in-law, were once heard talking on telephone in their Skardu language, that they should store enough grain for their maintenance. This Tehsildar of Leh was inciting hatred for the Indian Army, he was always non-co-operative towards our demands for supplies, transport and local resources.

ADMINISTRATION

Having realised the non-co-operative attitude of the Tehsildar, I contacted the locals directly for our requirement. I made Kalon Rigzin of Leh, who was then President of Young Buddhist Association responsible for providing recruits for the guerrilla force. He was also asked to get information about enemy agents and rumour-mongers and keep an eye on suspected local elements.

I appointed Shri Donyed as my supply officer to provide supplies, rations, fuel, mutton and vegetables directly from villagers and shopkeepers on cash payment at local rates. I appointed Mr Gulam Qadir, Jamadar of Leh Tehsil, to be responsible to provide transport, ponies and donkeys at local rates on cash payment. I also purchased cloth from the local market and employed tailors to stitch uniforms for the guerrilla force. We purchased local shoes of various types available in the market for issue to these men. I contacted the Zaildar and village head men and made them responsible for providing supplies and transport to our detachments. The soldiers were required to issue a receipt for supplies and transport which they received. The headmen were paid cash on producing these receipts to me or to my representatives. We also made use of the civil dispensary at Leh for looking after our patients.

I had a direct link on radio set with 19 Infantry Division at Srinagar and with Kargil and Skardu. There was also a single line telephone linking Kargil and Srinagar. In addition, we also organised a system of runners to carry messages by relay stages arranged by local headmen.

Training of our guerrillas was proceeding satisfactorily. The volunteers took keen interest and the morale of locals was also soaring high. They cultivated their fields and carried on their routine life without fear of the invaders.

I had two types of volunteers :

(a) Those who wished to be eventually enrolled in the regular army; and

(b) Those who wanted to serve just for the period of emergency.

The volunteers wishing to serve regularly were named Home Guards and those who wanted to serve only during the emergency were named National Guards. The Home Guards were issued a cap, two mazri shirts and two pairs of pyjama and a pair of shoes or boots. They were paid ration money @ Rs.2/- per day. The National Guards were not issued with clothing but they were paid Rs.2/- per day when on duty. No pay was fixed nor paid to them. They all seemed most contented and we had no trouble with them.

J & K TROOPS

Two platoons of J&K Infantry under Captain Badri Singh reached Leh. After a week's rest I despatched one platoon under Jamadar Harnam Singh to protect the bridge over Indus at Khalatse. This bridge was at the junction between the pony-track from Kargil and the track from Dah and Skyrbuchan along the banks of river Indus. The post was linked with telephone. The platoon commander was ordered to keep me informed about the situation.

Information was received from Nubra valley that there were movements noticed beyond Changmar. Some enemy troops were observed occupying high ground. The platoon at Deskit was, therefore, shifted to Changmar which is a holy place for the Buddhists and is located on the route coming from Baltistan.

Sensing danger to Nubra valley in April 1948, I despatched a platoon of irregulars under Subedar Bhim Chand to Nubra to reinforce the platoon at Changmar. Rinchen who belonged to Nubra valley also accompanied Subedar Bhim Chand.

On reaching Changmar, Subedar Bhim Chand appreciated the situation. He noticed enemy activities on a high hill feature stretching to approximately three miles. He led his men to this hill stealthily and attacked the post. The enemy was completely surprised and ran away leaving behind their arms and dead bodies. Our troops identified some of these troops as local Muslims armed with matchlock guns who were assisting the Pakistanis. Having accomplished this task, Subedar Bhim Chand and his party returned to Leh. They were given a rousing reception by the town's people and congratulated them on their victory. This was our first significant action. After some rest, I sent

Subedar Bhim Chand and his party back to Nubra to train the irregulars in the valley.

SITUATION AT KARGIL

5th Battalion of Jammu and Kashmir Infantry (State Force) commanded by Lieut. Colonel Kirpal Singh had arrived at Kargil. They had been ordered to move to Skardu and link up with the garrison and relieve them. One Major Coutts of Bihar Regiment was also sent as Indian Army liaison officer. Lieut. Colonel Sampuran Singh of 1 Sikh, with a platoon of Sikhs, was also at Kargil. He also had orders to advance to Skardu and link up with the garrison.

The first party which was sent to link up with Skardu garrison suffered heavy losses due to desertion of porters. These porters were mostly Muslims of Khanmarg area. The Government of Jammu & Kashmir therefore, decided to provide porters from the Ladakh region. I received a message from Mr.D.P. Dhar, Emergency Officer of Kashmir, to arrange for Buddhist porters and despatch them to Kargil. I promptly sent 100 porters and 50 ponies to Kargil.

I telephoned Lt. Colonel Kirpal Singh at Kargil that I had despatched porters to him. He enquired "Who told you to provide me porters. I don't want any porters. I have arranged porters here. The Raja of Khanmarg is providing me porters and he will accompany me." I told him that he should not rely on them, they may again deceive him, as they had done earlier. I added that I had sent the porters on the instructions of Shri D.P. Dhar. If he did not need these porters, he should send them back as they were already on their way to Kargil. When these porters approached Kargil they heard shooting and gunfire. They immediately stopped and two of them went forward to see what was happening in Kargil. They saw Pakistani troops doing a flag march in the Kargil town. These porters and ponies immediately retraced their steps to Leh. On the way they informed what had happened at Kargil to Jamadar Harnam Singh at Khalatse post.

KHALATSE POST

I was informed by Jamadar Harnam Singh that he had information that a Sikh Officer and about 16 Sikh soldiers were coming from Skyrbuchan to Khalatse, I had no idea who they were. I took it as a deception plan of the raiders in the guise of Sikhs to approach the post and then capture it. I, therefore, ordered him to cover their movement and allow only two men to enter the post and physically check them whether they were genuine Sikhs or Pakistani raiders. He did so, and then allowed Colonel Sampuran Singh and his party to enter the post and served them tea. I contacted Sampuran on telephone and advised him to stay on there. He flatly refused to do so, and said that he was not under my command and his men had no clothing, equipment and arms. I, therefore, told him to come to Leh.

Next day Major Coutts with Sixteen Gorkha Soldiers of the State Force also arrived at Khalatse. Since they were also ill-equipped I asked them to come to Leh. These two officers were not on good terms. Major Coutts told me that his signals operator, a Muslim Subedar, had deserted with the wireless set. We were still in communication with this set, so my wireless operator abused the Subedar in the dirtiest language and made no secret of his shock that the Subedar had turned traitor.

The morale and spirit of people in the Ladakh valley was quite high, but the news of the fall of Kargil and the arrival of Lt. Colonel Sampuran Singh and Major Coutts spread in the valley and the locals got agitated. The Muslim population in Leh felt very insecure on arrival of Sampuran Singh and his party. They requested me to protect them. I was well aware of the characteristics of the Muslims of Ladakh. I was very pleasantly surprised to see that the Muslims, Hindus, Christians and Buddhists were inter-related by marriage. The law in force was that the religion of children would be the religion of the father. They were more tolerant and human than what we are today. There was only one particular sect of Muslims who did not mix with the non-Muslims. I accommodated Lt. Colonel Sampuran Singh and his party inside the old fort and provided them all normal facilities. Major Coutts however stayed with me.

ORDERS FROM 19 DIVISION

I apprised the GOC 19 Infantry Division about the situation in Ladakh and the arrival of the survivors of the task force sent to reinforce Skardu. I was ordered to employ these 17 Sikh and 16 Gorkha soldiers for the defence of the airfield. They were soon deployed on piquets around the airfield.

ATTACK ON KHALATSE POST

I had positioned one platoon of Jammu and Kashmir State Force along with 40 irregulars at the bridge. I sent Major Khushal Chand to Khalatse to disburse money to the civilians who had provided supplies and transport to the troops. I had also instructed him to check the defences at Khalatse. While he was there, the enemy reached the other side of the bridge and fired at our positions on the bridge. During that night the whole platoon of the State Force deserted. Next morning, Major Khushal Chand alongwith the irregulars under heavy fire from the enemy personally sprayed kerosene oil and burnt the bridge. One splinter of 3-inch mortar bomb hit his eye, but he held the enemy at bay skillfully ambushing with the meagre strength he had under his command. These were mainly a few soldiers from 2 Dogra who were imparting arms training to the irregulars. There were unconfirmed reports from around Kargil that the raiders were now advancing in strength towards Leh. I sent a message to 19 Infantry Division that if reinforcements and supporting arms were not despatched to Leh it would be very difficult to defend Leh and the airfield.

I had to send a daily meteorological report to Srinagar. On completion of the airfield I had requested 19 Infantry

Division to send an aeroplane to Leh for a trial landing. Since no plane arrived a rumour spread that the Indian plane could not land at Leh, but the Pakistani planes would be able to do so. Since no plane arrived in Leh despite my best efforts, I sent a meteorological report on 22 May to the effect that all was clear in Leh and the weather was suitable for landing, although there were clouds that day. On the 21st night a rumour had spread that Pakistanis would arrive *at Leh on 22 May.* It was a very strong rumour. Next day on 22nd, almost the whole population of Buddhists and Hindus ran away to adjoining high hills and some of them were given shelter by their Muslim friends.

LANDING OF DAKOTA AT LEH : General Thimmayya and Air Commodore Mehr Singh

While I was still wondering how to handle the situation, I saw a Dakota landing at the airfield. I was riding a horse and so I galloped to the airfield. I stopped half way to check whether it was an Indian Dakota or a Pakistani. I saw Lamas of Spituk unloading some material from the Dakota and I also saw Mr Sonam Norbu and Babu Dorji standing there. On reaching the airfield I saw General Thimmayya and Air Commodore Mehr Singh talking to Norbu. I saluted the General. The General said, "Prithi, are you afraid of these Gilgit rats?" I said, "Sir, I am not afraid of them, but I have no regular troops. These irregulars are not yet fully trained. If you give me a battalion of regular troops, say Gorkhas, I give my word that not only will I chase these raiders but also will advance towards Skardu via Nubra and Chorbatla pass."

General Thimmayya responded, "You see the Dakota has landed. I will now send some troops at the earliest to reinforce you. But, you must defend this airfield at all cost." In the meantime Lt. Colonel Sampuran Singh also arrived. The General said, "Look Sampuran, your Battalion is without a commanding officer. You accompany me back." So he took Sampuran with him. Air Commodore Mehr Singh had already warned us that the weather was bad, and after ten minutes the plane would leave. The General asked me as to why he could see no movement of people in the Leh town. I told him that due to a rumour they had all run away to the hills that morning. I pointed to the hills from where the locals were now coming down having seen the Dakota landing. It is well-known that these two gallant senior officers had flown to Leh against orders from higher Headquarters. But for this timely disobedience we would have lost Leh.

MOVE TO KHALATSE

As I mentioned earlier, the entire platoon of Jamadar Harnam Singh at Khalatse had deserted. These men were now hiding in villages near Leh. When they saw a Dakota landing they got confidence and returned to the barracks. I collected them and gave them hell, abused them and told them that all of them could be tried by Court Martial for desertion. I wanted to put them under arrest, but since I had no troops to guard them, I decided to use them. After giving them a bit of my mind I asked them if they would accompany me to Khalatse. All of them volunteered. There was a rumour that some officers and troops of Pakistan had already reached Nemo. So, we first advanced towards Nemo. On my way I met Major Khushal Chand who had laid an ambush. I ordered him to return to Leh and have rest, while I along with the State Force platoon would advance to Khalatse and contact the enemy. Major Khushal being my real cousin, sensed danger to my life and said, "No, we won't go to Leh but advance to Khalatse. Instead, you should go back and control the situation at Leh." I tried to convince him that his injury did not permit him and that his men required rest, but he refused to obey my orders. To show my annoyance I whipped my horse and moved on. I could see my cousin and his men standing and watching me till I disappeared into the dead

The wounded soldiers looked after by Lama Sin Spituk Monastery in 1948.

ground. After a little distance an old man came to me and said, "You had promised to protect us but today you have thrown us to the wolves." I encouraged him and told him that since the Dakota had landed, troops will soon be sent by air. Thereafter, we resumed our move towards Saspol which was in the grip of severe panic with the rumour that Pakistani advance party was very close to Saspol. We were again stopped by the villagers who told us that the enemy troops were holding positions at Khalatse. That night we halted at Timusgang. Next morning we saw disorder amongst the villagers. Some of them had observed enemy movements on the opposite side of the river. We stayed for the whole day at the village and in the evening proceeded to village Skindiang via village Teya. At Skindiang we came to know that about four Gilgit Scouts soldiers and a team of masons and carpenters had camped in a house. They had constructed a raft at the bridge site and, while the main body of troops were on the other side of the bridge, they came to the bridge site daily and repaired the bridge. The villagers of Khalatse had been forced to work on the bridge.

At once I sent some villagers to Khalatse to tell the villagers to evacuate their village that night and come to Skindiang. The Indian planes would bombard the bridge site. At night I led my platoon to Khalatse where the enemy detachment was occupying a lone house. We surrounded the house and killed all the occupants. There were four artificers and three soldiers, one of the soldiers managed to run away towards Domkhar side. Then we went down to the bridge site, destroyed the raft and positioned a guard of 1 NCO and 3 OR to snipe at the enemy if they came down to the bridge for repairs. We returned to Skindiang village which is situated north of the Khalatse village. Next evening we relieved the post at the bridge. The Havildar in charge of the post informed me that they had shot one JCO and three other ranks, who had come there for repairs of the bridge. This Havildar and a L/NK who had shot the enemy had earlier deserted the post. That day, they were very proud of their action. Henceforth it became our routine practice to lay ambushes at the bridge at night and cover it with fire during the day. As a result, enemy stopped coming to the bridge site for repairs.

I also sent a few locals with a message to the Headman of the village Skyrbuchan to give me information about the enemy and their agents, if they had any knowledge about it. If such a person was identified, then, they should arrest and bring him to me. One night two middle-aged men of Skyrbuchan came to me and said that there were enemy representatives who were collecting rations for the enemy. The Headman got hold of them, tied them up and tried to bring them to me. But the men did not budge. So, the villagers threw them into the river Indus.

I was told that for the last two days, no movement of enemy was observed on the other side of the river. In order to check the sniper party and observe the front I went down to Khalatse. I saw our men in trenches, but there was no sign of the enemy. I then ordered the men to come out. All of us stood together to attract enemy fire and ascertain their presence, but the enemy did not fire. The troops again occupied their posts, and I and one civilian named Sherab walked back towards Khalatse. We hardly walked for about 100 yards when we were fired upon with LMG and rifles. The volley of fire was tremendous and we were caught in the middle of flat ground. I immediately told Sherab to take cover. He promptly responded by hiding his heal behind a rock with his body fully exposed to the enemy. This young man had rosy cheeks, but now he looked absolutely white with fright. I told him if we stayed there any longer we would be shot, so we must run away, but he was scared to come out. I told him that we could not be hit by those bullets, since we were Buddhists and our Gods would protect us. I ran and took shelter near a rock, so did Sherab. He was now in a dead ground and he signalled to me to run to him, but then I heard a bang of 3-inch mortar fire and to my horror, the shell burst a few feet from me. I had hit the ground and put my head in a hollow rock and left my body exposed to the mercy of God. The shells burst all around me. Enemy had fired approximately 10 shells. My body was covered with dust and I was still dazed when all of a sudden I heard my snipers fire. The enemy having sighted their position, switched over their fire towards the snipers, and I was spared. Thereafter, there was no fire on me. Taking advantage of the lull I ran towards the dead ground. Some shots were fired as I ran, but I was soon safe. I went down to the village where, Sherab and two other young men, on seeing me, immediately ran into their house. I thought perhaps Sherab having seen me covered with dust took me as dead, and now must have imagined me as a ghost. I was afraid that these young people might attack me but they came out from their houses with bottles of Arak (local alcohol) to welcome me. They greeted me with tears of joy and bowed their heads at my feet. I, in turn, embraced them and thanked them for their affection. We stayed at Skindiang for six days. During the day we remained inside the house. We could see enemy observation post on the other side. Each house had accommodated three to four families of Khalatse village. While the men and women went out to work in the field, I did baby-sitting for them.

I got restless doing baby-sitting with no signs of any aircraft landing at Leh with reinforcements. I thought the Division had failed to get Dakotas and, therefore, had probably written us off. Then at about 10 a.m. I noticed six Dakotas approaching Leh. I calculated that each Dakota would be lifting a platoon; hence approximately two rifle companies must have been flown. I sent a message to let them rest and acclimatise for three days. They should then be sent up to relieve me. But, to my dismay, it was a weak rifle company of 2/4 Gorkha Rifle of 77 men commanded by a Second Lieutenant with one year service. Lieut. Kaul was the son of a judge with no experience of warfare. He reported to me with his men and I handed over the Khalatse position with orders to cover the bridge and stop enemy advancing via Indus valley.

On my return to Leh, I had a real good bath and then met the 70-year old Meme Sonam, in charge of our residence. I told him to count of all the lice I had on my vest. After sometime Meme Sonam, scratching his head, came to me and told me that his eyesight was weak. He could count only the black and big lice but not the white and small ones. He counted 35 lice, and that was the decoration I had for this Khalatse operation awarded by the villagers of Khalatse and Skindiang!

2/4 GORKHA RIFLE

I received instruction from 19 Infantry Division that, instead of taking defensive action, I should now take offensive action. I, therefore, ordered Lieut. Kaul to despatch a small detachment to protect the Khalatse bridge and advance further down towards Skyrbuchan. He took his company about eight miles forward and then took a defensive position at Dumkhar. The enemy, on the opposite side, having seen our troops advance towards Skyrbuchan, also sent a company strength of Gilgit Scouts to Dah. The Company of 2/4 Gorkha Rifles in defence at Dumkhar was surprised by an enemy attack. They lost the high ground, left their position and withdrew to Hemis Shukpachen and took up position there.

2/8 GORKHA RIFLE

One Company of 2/8 Gorkha Rifles (called Chapati Column) under Major Hari Chand was sent on foot via Manali - Lahaul (Bara Lacha pass) to reinforce the Leh garrison. While they were on their way to Ladakh my generator set broke down, and inspite of all efforts I could not get the batteries charged. For almost a fortnight we had no wireless communication with the Division. I finally sent a team of four men on horses to contact the marching 2/8 Gorkha Rifles to get some batteries and return to Leh at the earliest. I soon got the batteries and was able to establish communication with the Division.

EVACUATION OF CIVILIANS WITH FAMILIES

At Leh there were about 40 Indian shopkeepers and several Kashmiri civilian Government servants with families. They felt that with the enemy advancing towards Leh it was best to leave. They requested me to provide them an escort. They planned to march to Lahaul via Bara Lacha pass. I told them that I was short of manpower, but if they wished I could give them a fortnight's training in arms and provide them a rifle each. They agreed and after a fortnight's training I made them fire two rounds on a target. They were able to hit the target and got confidence. I then handed over 20 rifles with 50 rounds per rifle and told them not to waste any ammunition. On arrival at Lahaul they should hand over the arms and ammunition to my brother Ex Captain Thakur Pratap Chand on a proper receipt. While leaving Ladakh they handed over the keys of their stores to me, with a request to use the goods for our Indian troops and to destroy the stores. These traders were mostly dealing with carpets, *'namdahs'* and Chinese silk from Yarkand and Sinkiang. They had large stocks of spices, tea and sugar to barter with the Yarkand traders. As I had to purchase many items from them, I detailed one Shadi Lal Goel, a matriculate trader to help me to provide stores from the shops and keep the account. I paid cash for the stores taken by me. My irregulars used *'namdahs'* as blankets and were issued silk and cotton shirts. The silk shirts were quite cheap and my men were in fancy colourful dresses. These traders also had stocks of Sinkiang carpets. Our Air Force pilots were greatly enamoured of these carpets. I used to stock these carpets by the side of the airfield, the pilots paid cash and purchased the carpets. These carpets then were valued at eight annas a square foot and the 'namdahs' at Rs.2 a piece.

SITUATION AT HEMIS SUKPACHAN

I was unhappy with the withdrawal of 2/4 Gorkha Rifles from Dumkhar to Hemis Sukpachan. I, therefore, went forward to check the defenses. I found that Lieut. Kaul had a single line defence with no depth. He never went forward and checked the positions on the ground. I ordered him to go forward and check the positions, specially the highest ground, and ensure that the enemy could not infiltrate and occupy the high ground above his position. He informed me that he was getting information from the post, but did not go forward to check it on the ground. One of the high features looked to be very vital. Lieut. Kaul confirmed that it was held by his men, but I doubted it. Later I came to know that this height was occupied during the day, only and at night the troops withdrew to a lower height, to save themselves from the effects of severe cold. Next day, while Lieut. Kaul was going up to check the position, enemy fired from this location. The enemy had infiltrated at night and occupied this high ground and dominated our positions. By now the 2/8 Gorkha Rifles company, under Major Hari Chand, had arrived at Leh after a long hard march of 42 days, having crossed three passes namely - Rohtang pass - 13,400 feet; Baralachala pass - 16,000 feet; and Tanglangla pass 14,000 feet. After two days rest, they came forward to Hemis Sukpachan and joined 2/4 Gorkha Rifles. Next day Major Hari Chand planned to raid the enemy Headquarters and thereby stop the enemy advance.

BATTLE AT HEMIS SUKPACHAN (SEE SKETCH 4)

Major Hari Chand and his company advanced at about 3 a.m. in the morning. At about 5.30 a.m. I went to the signallers and asked them to contact Major Hari Chand.

They informed me that their set was out of order for last few days, and they were not in communication with Major Hari Chand. While I was talking to them I saw 20 tall soldiers coming down from the hills towards the village. They were hardly 500 yards from me. I immediately realised that these tall men could not be Gorkhas. I immediately mustered the signallers, the cooks and the non-combatants and fired at these men. The enemy was surprised and took cover by entering the nearest village house, from where they fired back on us. I had again no troops to attack the enemy. In the meantime I saw about two platoons of enemy troops advancing towards our position to encircle us. One Naik Narpat of Gorkhas who was posted nearest to the village saw the enemy and engaged them with rifle fire. In the meanwhile, having heard the firing Major Hari Chand with his men returned and surrounded the house occupied by the enemy. A bullet hit Major Hari Chand on the arm but he remained undeterred. I told him to go back to Leh and get proper treatment. In his absence I offered to take command of his troops, but he refused. He said that since he had led his men, leaving them at that time would affect the morale of his troops. Moreover, since he did not understand Ladakhi language, he would not be able to get supplies from locals as I did. Since we had lost the vital ground and were surrounded, he requested me to return to Leh and arrange some air or ground support for them. I, therefore, ordered him to withdraw from that position and take up defence in a rear position.

Naik Lbonzang, an ex-Dogra sepoy of Shargul village, commanding a section of irregulars, was killed in this action. Three Gorkha Rifles sepoys were also killed and Major Hari Chand was wounded.

MONASTERY AT HEMIS SUKPACHAN

In the middle of Hemis Shukpa village on a high ground was a monastery. Inside this monastery the Lamas lived. When we had taken positions around the monastery and engaged the enemy with fire, the Lamas too started praying inside with their Puja music instruments for our victory. One of the Gorkha sepoys asked his Havaldar, *"Ustad Bhitar Lama harule ke kardaee chha"* (what are the Lamas doing there ?). Prompt came the reply, *"Hami haru ko lagi covering fire dia chha"* (they are giving us covering fire).

With a heavy heart, in the morning, I rode back from Hemis Sukpachan. I passed very close to the enemy post. God only knows why the enemy did not notice me. On my way back, I pleaded with the locals in the village to serve and help our troops, and they should not believe in rumours. I told them that I was going to Leh to arrange further re-inforcement of troops by air. While I was crossing the Phirjang nallah, the flooded nallah drifted me and my pony for a few yards. At about 2 am I arrived at Leh.

SITUATION AT LEH

When I entered the Leh Fort, Havildar Heshes Chhopel, an ex-Dogra sepoy of my irregulars informed me that the rumours had spread at Leh that myself and all the Gorkha troops were either killed or taken prisoner by the enemy at Hemis Sukpachan, and that the enemy was advancing towards Leh. So, Major Khushal Chand had shifted the Leh treasury to Martselang in case of withdrawal. I had already given orders to Major Khushal Chand who was suffering from an eye injury due to recochet of enemy bombing and also from hernia, to leave Ladakh by road for Lahaul for his treatment. The senior- most person at Leh was Havildar Quartermaster Ganga Bahadur Gurung of 2/8 Gorkha Rifles. When he came to know that the troops in front were killed and taken prisoner, he immediately collected my irregular Ladakhi troops and told them that in war sometimes we lose and sometimes we win. Now that we had lost and were not strong enough to face the enemy, we have to withdraw. But, before withdrawing, we must finish off all who were agents and sympathizers of the enemy. He posted men around Leh Bazar to stop people from sneaking away from Leh town, and tell them that it was all rumour and that I had arrived at Leh. On hearing of my arrival, Havildar Gurung said *that if the head was intact, the complete body would be intact.* He then came to me and saluted. He gazed at me with his blood red eyes. He could not address me at first, but then he said, "Sir, please have rest at night. I have planned everything for tomorrow's operations." When I told him that all officers and men of 2/8 Gorkha Rifles were alive holding their positions in front, he was pleased and seemed contented. I told him not to do anything without my orders. By then Major Khushal Chand had shifted the Leh treasury to Martselang for safety. Later on, the treasury was brought back to Leh. Major Khushal Chand had now left Ladakh for Lahaul for getting medical treatment. After treatment he was posted as CO of 14 East Punjab Militia. Mr. Devi Singh and Mr. Shiv Chand of Lahaul took a delegation to Delhi and requested Pandit Jawahar Lal Nehru to provide arms and give training to Lahaulis for the defence of Lahaul. Panditjee told the delegation that Major Prithi Chand, Khushal Chand and the brave men of Lahaul were fighting the enemy in Ladakh and hence there was no danger to Lahaul. This was a noble tribute to us who were fighting in Ladakh.

Major Hari Chand appreciated that the situation at Hemis Sukpachan had become untenable and withdrew his troops to Bazgo and ambushed the enemy advance and inflicted heavy casualties on them. The enemy pulled back to the hills overlooking the Gorkha position and Major Hari Chand once again had to withdraw to a safer place, from where he could stop infiltration and enemy advance. He took position on the Taru Ridge and deployed his men in defence. This position was *occupied on 6th July 1948 and remained in our hands till 18 November 1948* when we went on the counter-offensive after the capture of Zojila pass.

RAID ON ENEMY GUN
SEPTEMBER 1948

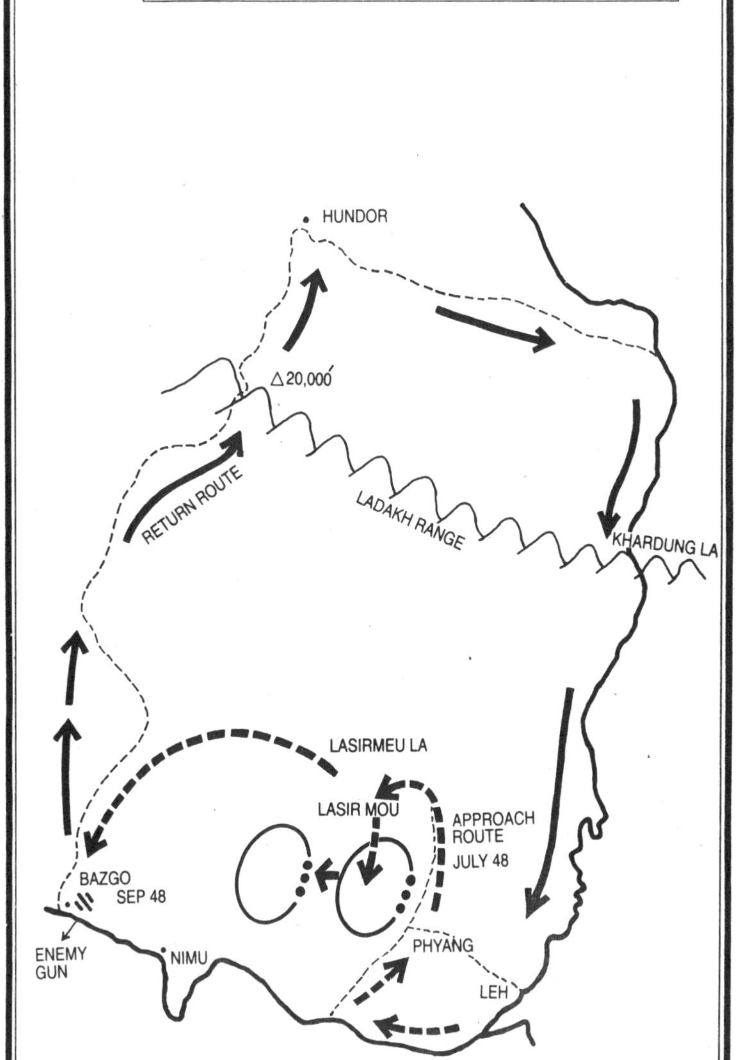

SKETCH - 5

ENEMY AGENTS AND FIFTH COLUMN

I came to know that some local agents normally sneak away at night from gaps in between our posts. I, therefore, employed my irregulars to lay ambushes on both sides of river Indus at night. We apprehended a large number of enemy agents trying to sneak away. In certain skirmishes, some of them were also shot by our men. I told the locals of the villages and the town of Leh that I would not spare those who were found to be sympathisers and agents of the enemy and who deliberately spread rumours. Thus, I was able to control rumours. While I was getting information from the enemy occupied area of Ladakh through my agents, the enemy did not get any worthwhile information from our side.

At Taru front the forward area extended from the hill tops to the river bed. It was not possible to hold it in depth, due to shortage of regular troops. *Consequently, I attached a section of irregulars with each section of Gorkhas.* Thus we raised a section to a platoon; and a platoon to a company; and a company to a two-company strength. Besides this measure, I had got the National Guard and civilians to hold certain less important areas. They were instructed to attract enemy attention with smoke during the day and with fires at night. Thus the enemy observed positions of our troops in depth. The enemy was also aware that we were getting regular reinforcements from Lahaul, and by air. They therefore, refrained from advancing to Leh. *The Divisional HQ nominated me as OC Leh Forces.* The enemy appreciated that at least a brigade strength was at Leh. *My designation as OC Leh Forces was later redesignated as OC Leh Garrison.* Since Major General Kulwant Singh objected that there was only one force commander, and that was he himself, there could not be two force commanders in J&K at one time.

DECEPTION IN NUBRA VALLEY

I had formed a Defence Committee of civilians in the Nubra valley to contain the enemy and deceive them. The members of the committee were Mr.Ganga Tsetan Phunthsog, the Naib Tehsildar, Mr.Kalon Lama and Gaga Rigzin Namgyal of Hunder. They, in consultation with Subedar Bhim Chand, formulated a plan for execution. The enemy at Shikhri could observe with binoculars our movements right up to Khalsar and beyond. *We used to despatch a fake convoy of 30 to 40 Yaks on every third or fourth day loaded with straw.* The enemy mistook this convoy to cater for a large strength of troops in Panamik area, and so they desisted from attacking our forward posts.

WITHDRAWAL OF SUBEDAR BHIM CHAND

Taru front was only eight miles from Leh. Since I had no reserves I had to withdraw troops from Nubra. It was an extended front. It was difficult to hold both the fronts at one time with limited troops. Since Taru front was more important, Subedar Bhim Chand and elements of the State Force were withdrawn from Nubra and were brought to Leh. In Nubra, only the irregulars under command of Chewang Rinchen (later Colonel Rinchen, double MVC and SM) with 20 rifles and 50 rounds of ammunition per rifle, were left to guard the valley. They divided their party in two groups. Mr.Thustob of Panamik, an ex-Serviceman of Gorkha Rifles, with five rifles and 20 others armed with sword and daggers patrolled the north bank of river Shyok above Udmaru village and Mr.Chhewang Rinchen with 15 rifles, and some others took position at Thoise on the south bank of the river, where there was a monastery at Skuroo. The defence committee as usual played their deception tactics well. The enemy was unaware of the withdrawal of regular troops from the front for almost a month. One day when an enemy patrol visited Udmaru in the evening, someone told them that there were no regular troops in Nubra, and that only locals with matchlock guns were deployed. Meanwhile Rinchen's party came down the hill and fired on them from the north. This gave an impression to enemy that the regulars armed with .303 rifles were in hiding nearby. More of the story of Nubra valley will be narrated by Colonel Rinchen in the next chapter.

RANI SAHIBA OF STOK

The Raja Sahib of Stok had seen the first Dakota landing at Leh. He came to me and requested to arrange an airlift from Srinagar to Leh for his son Kunzang Namgyal who was a student in a Baramula convent school. I, therefore, requested 19 Infantry Division to send this boy by air. When the Rani Sahiba came to know of this, she came to me and requested to cancel the proposed airlift for her son, instead, she wanted the boy despatched via Kulu-Lahaul-Ladakh road. This was arranged. Later when six Dakotas landed at Leh, the Rani Sahiba and others went to the airfield and were shown the plane and the troops coming out from the plane. After a few days she came to me again with a request to arrange an airlift for her son. I informed her that as she had wished I had already got it cancelled. Anyway, I enquired as to why had she changed her mind. She replied that she had never seen an aeroplane. *She thought that the passengers rode the aeroplane like a horse, therefore, was afraid lest her son would fall down.* But, since she had now seen that the plane was just like a house she was keen that her son should travel by air. But it was too late. Her son had already left Srinagar and was on his way marching with 2/8 Gorkha Rifles towards Ladakh.

The Raja of Stok is related to me. The present Rani, Parvati is my niece. Their son, Kunzang Namgyal on arrival in Ladakh was recruited in 7 Jammu and Kashmir Militia and got a commission. He has since died. Remembering the story of air travel I cannot help but recall a funny incident. *When the first aeroplane landed at Leh, a senior Lama of Spituk monastery approached the pilot and asked him as to what should he offer to the aeroplane: some food or a drink !.*

SUBEDAR BHIM CHAND

Subedar Bhim Chand had lost his wife leaving a two month old baby. I did not break the news to him, because I knew it would demoralise him. I had kept this news secret till one day when a Lahauli signal Havaldar who had accompanied Major Hari Chand broke the news. Subedar Bhim Chand cried and cursed me for not informing him about his beloved wife's death. I told him that our presence on this front was more important and I did not want to lose him since he had a high reputation in Ladakh. He was broken, and said that he had no guts left to fight, and requested me to sanction leave to go to Lahaul. He said that after arranging his domestic affairs, he would very soon return to Ladakh.

I informed Major Hari Chand that Subedar Bhim Chand had arrived from Nubra and the news of the death of his wife had given him a shock. Therefore, he had asked for leave. Major Hari Chand then requested me to send Bhim Chand to him so that he could get some information and advice about the enemy. I sent Subedar Bhim Chand with some respected civilians to Major Hari Chand. While the Subedar was on his way to Major Hari Chand, I told the latter that the refugees and village headmen from enemy occupied area who were employed as porters should surround Subedar Bhim Chand and tell him, that the army men from Lahaul had promised them to defend them from the invaders. Major Khushal Chand had gone, now if Subedar Bhim Chand also left, perhaps some day Colonel Prithi Chand will also retreat to Lahaul. But for the promise made by the Lahaul army men, the refugees would not have stayed in their homes. Since he had lost his wife, a girl of his choice would be provided to him for marriage, but he should not leave.

This plan worked. Subedar Bhim Chand was persuaded by the refugees to stay and help Major Hari Chand to fight

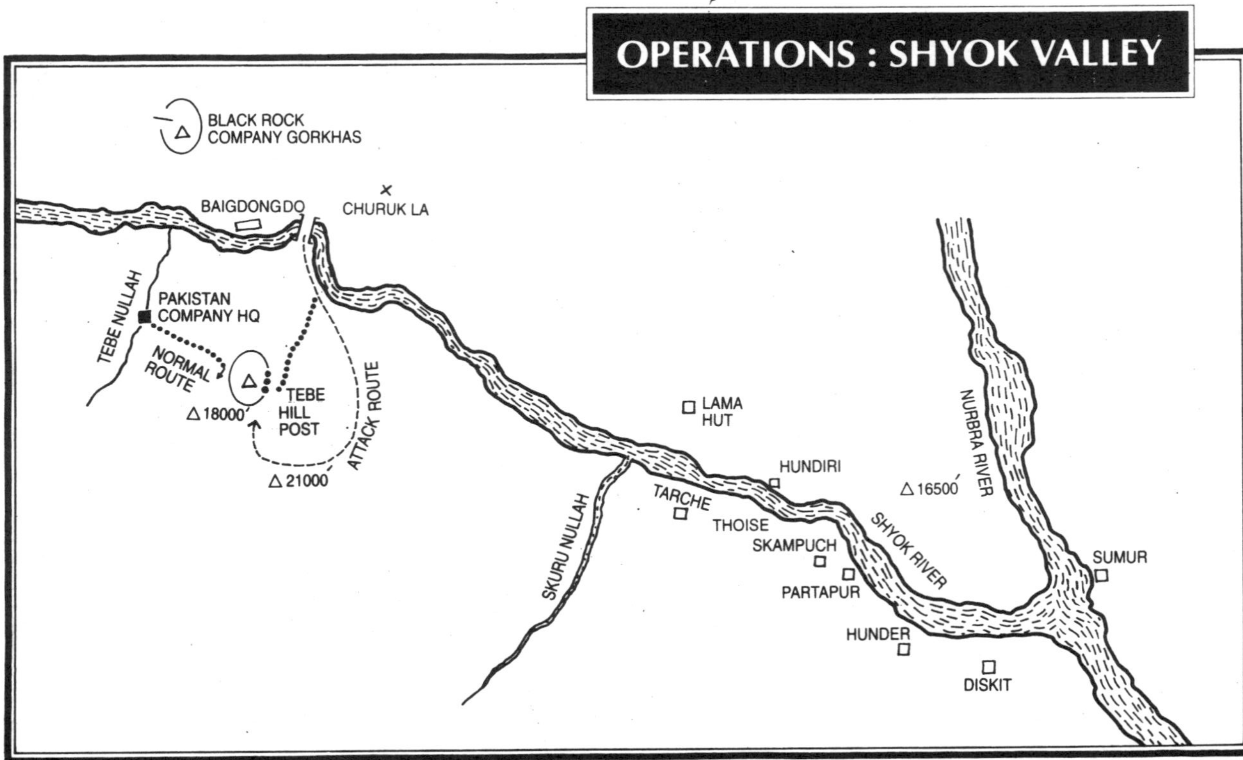
OPERATIONS : SHYOK VALLEY
BLACK ROCK
COMPANY GORKHAS
BAIGDONGDO
CHURUK LA
TEBE NULLAH
PAKISTAN
COMPANY HQ
NORMAL
ROUTE
△ 18000'
TEBE
HILL
POST
ATTACK ROUTE
△ 21000
SKURU NULLAH
LAMA
HUT
TARCHE
THOISE
HUNDIRI
SKAMPUCH
PARTAPUR
HUNDER
SHYOK RIVER
△ 16500'
NURBRA RIVER
SUMUR
DISKIT

SKETCH - 6

the enemy. Major Hari Chand also said that if he was in the same position he would not leave the front but do his duty to protect the honour of the inhabitants of Ladakh. After weeping for sometime Subedar Bhim Chand told Major Hari Chand that he preferred to stay on and kill the enemy. A little later Subedar Bhim Chand spoke to me on telephone and requested me to send his irregulars, his revolver and his batman since he was joining Major Hari Chand for the attack on Pherjang.

BATTLE OF PHERJANG

A platoon of Gilgit Scouts had occupied this dominating vital ground known as Pherjang hill. It was very important to capture it. Therefore, in the early morning, this post was attacked and the position was captured. Major Hari Chand chased the enemy further and again captured three more important posts. Later, these posts were occupied by 2/8 Gorkha Rifles and Subedar Bhim Chand and his men returned to barracks in Leh. In this attack Sepoy Tobge Ram of 2 Dogra did very well. He got a Vir Chakra for his gallantry in this action. After this action Subedar Bhim Chand also left for Lahaul to settle his domestic affairs.

SITUATION IN LADAKH IN AUGUST 1948

Mr. Chhewang Rinchen and his irregulars were holding Nubra valley front with intensive patrolling. Major Hari Chand with his 2/8 Gorkha Rifles company, a company of 2/4 Gorkha Rifles commanded by Lieutenant Kaul and about 100 irregulars was holding piquets on the Leh front. The troops in contact had only their pouch ammunition. There was no reserve ammunition. We did not have any sort of medical aid. I had detailed a local *Amchi* (Tibetan Ayurvedic doctor) to treat our patients. The stock of rice and wheat was exhausted. I started to issue 'Sattoo' to the troops. Sattoo is made of roasted barley ground as flour. It was boiled in hot water and made into a pudding like 'Halwa'. It was then issued instead of Chapaties. It is eaten with dal or meat or vegetable curry. At first the Gorkhas who had fought in Africa and Italy and had good rations there, threw away the Sattoo balls saying, "Is it the Indian army ration". But, our irregulars relished it. Later, when the Johnnies felt hungry they asked the Ladakhi boys how to eat the 'Sattoo'. Once the Johnnies got a taste of it, they demanded Sattoo pudding balls instead of parathas and chapaties because it was soft to eat. Cooked food was sent to picquets from the base. Hence, by the time these parathas and chapaties reached the high piquets these became hard like a cardboard. Similarly troops were supplied with locally made woollen socks and shoes of all sorts including boots with wooden soles available in local market.

MORALE OF TROOPS

The troops were quite happy with ration supplies but shortage of ammunition and lack of medical aid affected the morale of troops. The morale of civilians was equally high. Parents, wives and other relatives were permitted to visit piquets to see their men. There were bunkers for visitors in each picquet. The visitors often offered them home made cakes, *chang* (local beer) and Arak (local alcohol). The soldiers too were permitted to visit villages in turn. The civilians while on piquets for visits often had the experience of being under the enemy shelling and firing. In the field they could make out the type of firing being resorted to on the piquets. The brave Ladakhis who had fought many battles against Mongols, Chengez Khan and Tibetans were once defeated and ill-treated by General Zorawar Singh's Army. They were once again confident of being victorious. Irregular Ladakhi Jawans had defeated the enemy on the high hills and destroyed many ammunition dumps behind the enemy lines. The enemy on several occasions had attacked our posts at night but each time their attack was repulsed by these brave Ladakhis. However, the morale of troops received a set-back on 14th

August when the news of the surrender of the Skardu garrison was received.

FALL OF SKARDU GARRISON

One morning my wireless operator called me on the set. Lieutenant Colonel Sher Jung Thapa, Commander of Skardu Garrison, wanted to speak to me. He said that he was sending this message through my set, because he could not directly contact 19 Division. He was soon able to communicate with Lt. Col. Sri Ram Oberoi of 19 Infantry Division. In the end, Lieutenant Colonel Thapa told Col. Oberoi, "At this last moment I request you and the General to excuse me if I had been rude or annoyed you in my messages." Then he said to me, "Prithi, I do not know whether I will ever see my family at Dharamsala. Please go there personally and convey my good bye to them." After this conversation the signaller gave his home address to the Division signaller and requested him to inform his family of their fate. Having said so Lieutenant Colonel Sher Jung Thapa destroyed the set.

In 1951, when I came to know that Colonel Thapa had arrived at Jammu, I went to Jammu and met him. For the first few minutes we could not speak due to emotions. Tears rolled down over cheeks and then we talked about our operations at length from 9 a.m. to 3 p.m. till one of the officers said, "Sir, the lunch is getting cold." I am really proud that we have such eminent men like Brigadier Thapa in the Army who took all the risks and troubles to save the lives of the refugees. He considered it inhuman to leave them in the fort when he could have easily retreated to Ladakh with his men. His patience, selfless sacrifice, human outlook quite rightly made him eligible for the highest military decoration. Brigadier Sher Jung Thapa, MVC returned to active service, and after retirement lives in Dharmasala, a little distance from the abode of His Holiness the Dalai Lama.

15TH AUGUST : Independence Day

The downfall of Skardu gave me a shock. I sent a message to 19 Division that the episode of Skardu should be an eye-opener, and if we were not supplied with medicines and ammunition, we too may face the same fate. Since June, after the landing of six Dakotas, no plane had arrived. I had kept all wounded soldiers and sick patients at Spituk monastery. I had no doctors and nursing staff to attend to them, so I had to keep them under the care of the Lamas. I made arrangements to brew our own wine, which I supplied to the patients in the monastery.

Major Hari Chand and I then planned to celebrate our independence day by attacking a number of posts by simulating noises, thus drawing fire from the enemy. The enemy retaliated by active firing while we conserved expenditure of ammunition. The locals at Leh organised a tea party for our men to celebrate the first Independence day.

ARRIVAL OF A MEDICAL OFFICER & AMMUNITION

On the 16th morning we got our doctor with medicines and ammunition by air. An Indian Army Doctor Captain Burdhan from Bihar landed at the airfield. We offered him a horse to ride, but he was reluctant due to the fear of falling down. This was the only transport we had, hence we taught him how to ride. Later, the doctor became a good rider and often demanded a horse to ride. On receiving the ammunition Major Hari Chand with his troops carried out many attacks on the enemy posts. The intense volume of fire not only demoralised the enemy but on certain occasions he vacated his post without giving a fight.

VISIT OF COL. B.M. KAUL (Later Lt. Gen. B.M. Kaul)

We now started getting supplies occasionally by air. One day Colonel B.M. Kaul (later Lt. Gen. B.M. Kaul) landed at Leh and told me that he was nominated as the Commander of Jammu and Kashmir Militia force. He informed me that the irregulars in Ladakh henceforth should be mustered in the Militia force and will be called 7 Jammu and Kashmir Militia Battalion and that I was the commanding officer of this Battalion in the rank of

Changla Pass 18,380 ft; Colonel Prithi Chand's column in March 1948.

Lieutenant Colonel. So far I held the rank of a local Lieutenant Colonel. Colonel Kaul had to stay for five days in Leh because there was no aircraft to take him back. Colonel Kaul visited the forward troops, appreciated the need for logistic build up and reviewed the quantum of troops required for the operations. He came to the conclusion that if we had to hold Ladakh, reinforcement of at least a battalion was urgently required. Proper replenishment of ammunition, minimum supporting arms and adequate supplies were necessary to fight effectively.

Colonel Kaul stayed with me in the Leh Fort. At night while I was fast asleep, having had my daily quota of Arak before sleep, Colonel Kaul a teetotaller spent a restless night killing bugs. Next day he helped me in writing citations for the officers and men for their acts of gallantry. He stated that he was carrying samples of our meagre food, clothing and shoes to Major General Thimmayya to see our awful condition. He promised that he would either arrange the evacuation of our troops or send reinforcements. Colonel Kaul then left Leh and soon we received a message from him that essential items, commodities and reinforcements were being released on priority. I came to know that he had flown to Delhi and apprised the Prime Minister, Pandit Jawahar Lal Nehru. I must thank him for all he did to help us at that critical time.

MILITARY GOVERNOR OF LADAKH

Lieutenant Colonel H.S. Parab, Commanding Officer of 2/8 Gorkha Rifles, landed in Leh in the third week of August with a platoon of Gorkhas. He told me that the remaining Battalion was marching on foot from Lahaul via Bara Lachala. He said that he had been detailed as Governor of Ladakh and commander of all troops in Ladakh. I was now relieved of the responsibility of being an overall commander. Henceforth I was to be Commanding Officer of 7 Jammu and Kashmir Militia. On his query as to how I managed the administration of supplies etc. in Ladakh and in the Nubra valley, I told him that I dealt directly with the locals through my civilian friends. I also informed him that the senior Tehsildar was unreliable and non-cooperative. Although most of the Muslims were in our favour, a few were fanatics and agents of Pakistan, who spread unwarranted rumours. He then asked me to arrange a Darbar to be attended by everyone. Next day the Darbar was held which was attended by all troops and civilians of Leh.

MILITARY GOVERNOR AND HIS MINISTERS

A big crowd had assembled. On arrival of Col. Parab all stood up and received him with 'Jai Hind'. Colonel Parab addressed them as follows :

"Ladies and gentlemen. I have been appointed as the Governor of Ladakh by the orders of Indian and J&K Governments for the administration and defence of Ladakh. The people of Ladakh from now on must obey my orders and give all assistance to me to carryout my task. Anyone who does not obey me or who is found in sympathy with enemy will be punished," and while saying so he took out his Khukri from the sheath and banged it on the table. He said he had nominated his Ministers, and everyone should co-operate with them and help them. He then announced his Ministers as follows :

Lt. Colonel Parab	—	Governor
Rev. Driver of Morravian Mission	—	Chief Minister
Mrs Driver of Moravian Mission	—	Health
Khawaja Abdullah Saheb	—	Armament
Mr Sonam Norbu	—	Engineering &Works
Mr Kalon Rigzin	—	Defence
Shri Donyod	—	Supplies
Shri Gulam Rasool	—	Transport

Having nominated his ministry he directed them to find their staff and carry out his orders. He did not consult me about his selection of Ministers. Rev. Driver, a British priest of Moravian Mission was often heard propagating for Pakistan. He had said that the Indian Dakotas were not capable of landing in Ladakh, because they could not fly high, but the Pakistani aeroplanes could enter Ladakh by flying low via the Indus valley. I had even informed 19 Division about him and his propaganda. When he was nominated as the Chief Minister I thought that Sheikh Abdullah and General Thimmayya must have been consulted and told Parab to entrust Rev Driver certain responsibilities. Later when General Thimmayya paid us a visit on 29 August he asked me as to why Reverend Driver was chosen the Chief Minister; I had earlier reported about his activities against our Government. I told General Thimmayya that I was not consulted in this matter. The General, therefore, ordered Colonel Parab to relieve him from duty. After the departure of General Thimmayya, Colonel Parab called me and said, "Go and tell Rev. Driver that he is no more the Chief Minister," and he ordered me to take over the duties of Chief Minister. I had to break this news tactfully to Reverend Driver. He was, however, very pleased to hear that he had been relieved from the post.

VISIT OF MAJOR GEN. THIMMAYYA

It was on 29 August that General Thimmayya paid us a visit. He seemed to be so tired that he was feeling sleepy. He had been working hard beyond normal hours to carry out his duties at Srinagar. In Ladakh there were no telephone calls and messages to disturb him. Hence, he was relieved and felt sleepy. After an early dinner he went to bed. Next day a message was received that due to bad weather no plane would fly to Ladakh. I told the General and he was pleased to hear of it. While he was having breakfast in a relaxed mood, Captain Badri Singh told him that a Dakota had just landed at the airfield. The General immediately galloped on horse-back to the airfield. There, Laloo Garewal, the Pilot, was waiting for the General. The General said he had received a message from his Headquarters that there won't be any flight from Leh. How was it that he had landed. Laloo said that according to the weather report he was not supposed to fly, but since the General had told him the previous day to pick him up, he was duty-bound to keep his promise, and had come to pick him up. These brave pilots took great risks and kept the air-link to Leh open at all times. I have already mentioned about Air Commodore Mehr Singh. This was the second incident that proved the dare devil spirit of our air force pilots. I attribute a lot to these air force pilots who saved us in Ladakh. God bless them.

REINFORCEMENT TO NUBRA VALLEY

One platoon of 2/8 Gorkha Rifles which had accompanied Lieutenant Colonel Parab by air, from Ferozepur Cantonment, was detailed for operations in the Nubra. I was told to guide them. This platoon which came from Punjab had no acclimatization at these heights and the low temperature. We marched to Ganglas village at the foot of Khardungla pass, which is 18,500 feet high, and spent the night there. Next day, we left at 4 a.m. for Khardung village across the pass. I provided ponies to the troops. We reached Khardung village after crossing the Khardungla at about 10.30 a.m. The last man of this platoon reached at about 9 p.m. Due to sickness, two of the Gorkhas had to return to Leh. At Khardung, dinner was provided but the men refused to eat because they were all suffering from headache and vomiting due to the height. Subedar Rana was depressed and wondered as to how they could fight the enemy with the rather poor health of his men. I consoled him that the troops will soon be acclimatized and be fighting fit. I then ordered him to issue a tot of rum to his men. The rum worked well and they ate their food. Next day I found them in good shape. I led them up to village Deksit and returned to Leh.

ARRIVAL OF 2/8 GORKHA RIFLES AT LEH

The battalion under the command of the 2 I/C Major Maini arrived at Leh after a march of 150 miles from Lahaul. They were given a good rest, and thereafter sent forward to relieve the men of 2/4 Gorkha Rifles Company, 2/8 Gorkha Rifles company and two platoons of J&K State Force. My irregulars who were now soldiers of 7 J&K Militia stayed on in the front. One day I went to the front to see my men, I was surprised to find that some men who should have been on duty in the front were at Leh. They said that they took permission of the piquet commanders and in their place they were relieved by their father or brother or cousin in the front to get some rest at home. I immediately sent them back and told them that they were now regular soldiers of 7 J&K Militia under my command and without my orders they could not leave their posts nor could they be relieved by their relatives in the future.

ENEMY PLAN OF ATTACK

Zojila pass had been captured by Pakistan troops in May. Having fortified their defences, one officer, Captain Sher Ali, with full company's strength was sent towards Ladakh for a final attack on Leh. The Maulvis had forecast that the stars were in their favour till 6 July 1948 for the capture of Leh. In Ladakh, the Lamas had told me that the stars were not in our favour till 6 July, but thereafter there was no danger. While Captain Sher Ali with his troops arrived at Kargil the Muslim representatives of Padam village from Zanskar valley told them that the Indian Army was advancing from Lahaul to Zansakr valley; and if not stopped, they would advance to Kargil. Appreciating the situation Captain Ali planned to first stop the advance of

Indian troops and later proceed to Leh. He therefore advanced to Zanskar valley and thus Leh was spared.

INDIAN ARMY IN ZANSKAR VALLEY

The Buddhist population of Zanskar valley having heard the news of the arrival of the raiders and other types of rumours spread by some Muslims of Padam village, were frightened. Some representatives visited Lahaul and asked my elder brother ex-Captain and formerly the Prime Minister of Lahaul Thakur Partap Chand for help. My brother was in possession of 20 rifles with ammunition, which I had provided to the Indian traders for their protection. At this time, Subedar Bhim Chand was also on leave in his village. My brother called Subedar Bhim Chand and discussed the matter with him. The rumour was that the Pakistanis were advancing to the Zanskar valley with the aim of capturing the Baralachala pass, and thereby cut off the line of communication to Ladakh garrison. Sensing the danger, my brother, Subedar Bhim Chand and Thakur Nihal Chand (brother of Lt. Colonel Khushal Chand) decided to go to Zanskar with the ex-servicemen of Lahaul equipped and armed with these 20 rifles and some private arms. They advanced to Zanskar valley and went forward to Padam village. At Padam they encountered the enemy and had a whole day's fight. They were only 30 men against an enemy rifle company with LMGs and mortars. Our men were encircled, and at great risk they retreated at night and made for Lahaul. They were chased for a few miles. This was the first battle fought in the Zanskar valley. The other battle was fought in December with a platoon of the enemy at Padam village. 77 Para Brigade while advancing to Kargil after capturing Zojila and having established a link with the troops in Ladakh at Kargil, sent a column to Zanskar. They surrounded this village and ordered a surrender. In the meantime there was a ceasefire on both sides on 1 Jan 1949. Hence the Pakistani troops remained at Padam. Later in March 1949, Subedar Murtaza of Gilgit Scouts from the Pakistani side was sent to Padam. He convinced the Pakistani troops at Padam that the ceasefire had been ordered and they should surrender. General Thimmayya later asked me as to how Pakistani troops got to Zanskar valley. When I told him the whole story, he had a good laugh. Thus the plan of enemy's offensive on Leh was diverted towards Zanskar.

DESTRUCTION OF ENEMY ARTILLERY

We got information from enemy occupied area that a big gun which was carried by 103 men was being brought by the enemy. We guessed this could be a 3.7' *mountain gun* and its ammunition. The enemy commander was Lieutenant Colonel A.G. Khan of Gilgit. He was a gunner in J&K State Force and had been in Africa during the Second World war. We appreciated that Lieutenant Colonel Khan had sent for the mountain artillery to support his operations. On the other hand, we had only four 3' Mortars with a limited number of bombs. The 3.7' gun was positioned at a location from where they could easily engage Leh, with its long range. A raid party under command of Major Hari Chand was formed. This party consisted of Subedar Bhim Chand and selected men of 7 J & K Militia, and some Gorkha and Dogra soldiers. In all there were 30 men. They were equipped with their personal arms, pouch ammunition and grenades. Limited supplies were carried by their men. We were not able to provide a radio set to the party - hence they were ordered to destroy the enemy gun and return to the base. The party left for Phiyang from where they went north up to Lasermola, turned west and then descended to Bazgo at night. The party consisted of a few young men of 7 J&K Militia who belonged to the village Bazgo. The villagers led the raid party to the gun position at night, which was by the roadside. But the crew of 10 Pakistanis were sleeping in a house near the gun position. The villagers also informed that a proper gun position had been prepared further ahead where the gun was proposed to be shifted the next day. Leh would then be within easy range of the gun. Major Hari Chand and his party went to the gun position at about 2 am and threw grenades inside the room and killed the entire crew. They had no explosives except grenades. They showed grenades inside the barrel of the gun and damaged the barrel. They also piled fire wood around the gun and burnt it. After having a good meal and having filled their pockets with bread and 'sattoo' they retreated. They crossed the 17,782 feet high Lasirmola pass and came out in the Shyok valley. From there they returned to Leh via Khardungla after taking a long detour. The enemy was visibly shocked by this bold exploit. Some locals of Bazgo village had informed the enemy that the raid party consisted of a few men of the village. The Pakistanis took revenge by shooting all male members of those families. When we received this news, we kept it a secret and did not reveal it to the affected men. They came to know only while chasing the retreating Pakistanis. On hearing the tragedy in their families, these men instead of losing their heart chased the Pakistanis with greater vigour (Refer Map 2).

RAID ON ENEMY AMMUNITION DUMP

My agents in the enemy occupied area informed me that the Pakistan troops had stocked ammunition, clothing and ration at Lamayuru rest house. I again sent a raiding party under Major Hari Chand and Subedar Bhim Chand via Chilling Sumdo on the southern side of river Indus. The raid party advanced from peak to peak, and after a march of four days they arrived at Lamayuru. They contacted the local headman and observed that only a section of enemy troops was posted for the security of this dump. In the evening, when it was dark, the enemy was sitting by the

Beasts of burden used to carry supplies to Shyok Valley in March 1948. The double hump Camel is in the back ground and the Yak is in the foreground.

side of a bonfire and gossiping. The raid party surrounded the rest house and opened fire with machine gun at close range and killed all of them. After that they poured kerosene oil and burnt the dump and destroyed the complete stocks. They waited that night for enemy retaliation, but there was none. They, therefore, extricated themselves the next day. The village headman knew that the enemy would commit atrocities on the villagers. So he sent his younger brother to Khalatse to inform Colonel of Pakistan Army with a request not to visit Lamayuru because the Indian troops were there. We came to know later that the Colonel immediately rode his horse and ran away towards Skyrbuchan.

Our raid party then withdrew along the same route and reached Leh safely. At Leh this party was paraded in front of Leh Fort and were given a heroes' welcome. They were dressed in Pakistani uniform, padded cotton coats and long Chinese boots. They even had Pakistani medals on their chests. These brave men were awarded cash prizes and given a weeks leave. They were also felicitated by the locals of Leh.

DEPLOYMENT IN THE INDUS VALLEY

At the end of October, the following troops were deployed:

Taru Front

One company 2/4 Gorkha Rifles, two companies of 2/8 Gorkha Rifles, one company of 6 J&K State infantry and about one platoon of 7 J & K Militia with tactical Headquarters.

Chilling Sumdo Front

One company of 2/8 Gorkha Rifles, two platoon of 7 J & K Militia. The Chilling Sumdo position was on the south of river Indus and on the east of Zanskar River.

Leh

Rear Headquarters, company of 2/8 Gorkha Rifles, Headquarters 7 J & K Militia along with training centre consisting of 12 platoons.

Supply Depot

RAP Hospital

Shyok Valley

One company of 2/8 Gorkha Rifles

One Platoon 7 J & K Militia.

STAFF FOR 7 J & K MILITIA

7 J & K Militia, on being raised had no staff. Since a large number of documents for enrolment had to be prepared and 'A' and 'Q' branches had to function, there was an urgent need to recruit trained staff to the battalion. I could neither recruit matriculate clerks nor employ civilians on an ad hoc basis, since there was an acute shortage of these categories of personnel in Ladakh. I, therefore, requested General Thimmayya to grant me a few days leave, so that I could visit my family in Dharamsala, and also recruit clerks for my battalion. Besides, I had to make arrangements for payment of cash to my wife, since my bank account was left at Rawalpindi. General Thimmayya reluctantly granted me 28 days leave but cautioned me that I may be recalled at any time. At Srinagar, I was surrounded by press representatives. Mr.R.K.Karanjia of "Blitz" invited me for a lunch and I gave him some unclassified information about Ladakh. Later,

after my return from leave I came to know that Mr.Karanjia had published an account of me and my men in the "Blitz" magazine. In the front page my photograph appeared describing me as the "The deathless man" for having crossed several high passes in winters under extreme climatic conditions with no snow clothing or equipment.

MY FAMILY

I came to Dharamsala and gave a surprise visit to my wife. For long she was unaware of my whereabouts. Many a time she had heard rumours that I and my men were killed or taken prisoners but she was confident that nothing would ever go wrong with me. She had seen 'Durga Mata' in her dream who told her that she was protecting me and she need not worry. I do believe that there is a certain amount of truth in this belief since on many occasions I was saved from death by incidents which were very close to miracle.

RECRUITMENT OF STAFF FOR 7 J & K MILITIA

I approached Sardar Kapur Singh, Commissioner of Kangra, to help me in getting clerks for my unit. He very kindly released some of the ex-servicemen staff from his organisation. I therefore, got a Jamadar Head Clerk, a Havaldar Quartermaster clerk, an Adjutant office clerk and two other clerks. I brought these clerks to Pathankot with me and flew them in a Dakota to Leh. These clerks were of middle age and hence I had to take care of them during the flight. There was shortage of oxygen since the plane had to fly high. When we landed at Leh and gave a good drink and dinner to my staff, one of them asked me as to why I had been often looking at them during the flight. I told him that I was scared of the height and their age. They seemed happy to see my concern for them.

DOCUMENTATION OF 7 J & K MILITIA PERSONNEL

A list of all serving irregulars and those who had taken part in the operations was prepared. Their documents for recruitment, medical examinations and battalion Part I and Part II orders were published. The enslistment of the irregulars commenced in April 1948, but a number of these irregulars were evacuated with battle wounds and were in Military Hospitals in Delhi and Srinagar. On being discharged from the Hospital in Delhi, these irregulars were despatched by train to Pathankot. Due to improper reception and guidance facility at Pathankot some of them were lost in transit. Those who were either at Srinagar or were able to reach Srinagar on their own were mustered by J&K Militia Headquarters at Srinagar and directed to Leh. Yet some of them never reached me. A few of them reached Amritsar where they visited Harmander Sahib. After praying they came out, and ignited a Beedi. A sewadar saw them and turned them out of the Gurdwara. These innocent men did not even know that smoking was prohibited in Gurudwaras.

JAMADAR CHHEWANG RINCHEN

Chhewang Rinchen of Nubra valley was a student of class 8th in Leh school. Rinchen turned out to be an inspiring leader. He was a fearless man and highly patriotic. He volunteered to take part in several battles and raids. He was recommended for his gallant deeds, and was awarded Mahavir Chakra. Later, he was commissioned as a Lieutenant in J & K Militia. 7 and 14 J & K Militia were later reorganised into the Ladakh Scouts. Lieut. Rinchen again proved his worth by his gallant actions in the 1965 war and was awarded the Sena Medal. He was again awarded the Maha Vir Chakra in 1971 war with Pakistan. He has since retired as a Colonel and now settled in Leh. A more detailed story of this stalwart is in the next chapter.

In November, 1977, the Parachute Brigade executed the plan masterminded and pushed by General Thimmayya to save Ladakh. The Brigade recaptured Zojila pass and then rapidly advanced to Kargil. Thus, the major route of

Pakistani attackers to Ladakh was severed. The rag-tag defenders of Leh went on the offensive and linked up with 77 Brigade. At last, Ladakh was rescued after being on the verge of falling into Pakistani hands.

CHAPTER 6

Colonel Chhewang Rinchen, Double MVC, SM, and Nubra Valley

"In August, 1948 the enemy had thrown all his reserves to capture the Nubra Valley. Naib Subedar Chhewang Rinchen with only 28 untrained National Guards held the enemy at Skuru Nullah for 23 days.

In September, he was detailed to capture the enemy position at Lama House; this was an extremely difficult task and entailed four days of march through treacherous country, including crossing a mountain feature over 17000 feet. He succeeded in capturing the objective with heavy casualties to the enemy and the capture of 13 rifles and one stengun.

On 15 December, 1948, when ordered to capture a hill feature near Biagdangdo he walked through snow for three days and succeeded in forcing the enemy to withdraw.

Again on 22 December, 1948, he was detailed to attack the enemy's last position in Leh Tehsil area. It took him six days to reach his objective. He had to go over a mountain feature 21000 ft and though his platoon suffered 50 percent casualties from frost-bite, he kept his men going through his outstanding and exemplary leadership. He attacked the two enemy posts and captured them; the enemy suffered heavy casualties.

This JCO displayed exemplary courage, inspiring leadership, initiative and the ability to plan and carry out his schemes successfully under most adverse conditions."

—Citation for Maha Vir Chakra

Of all the achievements of this seventeen-year old youth the most decisive was his stand for 23 days on the Skuru Nullah. It was the time when the enemy advancing on the main route to Leh had already reached about ten miles short of the town. Had the enemy broken through the Nubra Valley and attacked Leh from the rear the story of Ladakh might well have been different. Highest and genuine tribute in war is always the one which is paid by the enemy. In 1984 Colonel Mohmad Yusuf Abadi (retired) published his memoir in Urdu, in a book titled *Baltistan Par Ek Nazar* (A Glimpse of Baltistan). The Colonel was the Pakistani commander leading the Gilgit Scouts in 1948. He says - "Our intelligence revealed that our repeated attacks on Skuru were foiled by the personal valour of a 17-year old boy named Chhewang Rinchen .... had we succeeded at Skuru, there would then have been no real obstacle to our capturing Leh."

What is the story of this man who became a legend, indeed a romantic war hero at the age of 17 years ?

I first met Rinchen in 1968. He was commanding a company in 14 Jammu and Kashmir Rifles in my Brigade. In a two-sided exercise I noticed the uncanny mountain sense he displayed while moving for an attack on an enemy picquet. He had God-given instinct to choose the most appropriate, even though the most hazardous route, to surprise the very vigilant enemy. I realised his being a man who comes into his own in battles. Little wonder that he is the youngest winner of the Maha Vir Chakra in the Indian Army, or for that matter in nearly two hundred years of history of the British Indian Army (Maha Vir Chakra being equal to the second highest gallantry award after Victoria Cross, the distinguished Service Order or Indian Order of Merit). Little wonder that he went on to win a Sena Medal in 1962 in the war against the Chinese and a second Maha Vir Chakra in 1971 in the war against Pakistan.

It is best to read his story in his own words :

CHILDHOOD

I belong to a celebrated warrior family in our region. Our house is named Stakre which in Ladakhi means 'lion'. This title was conferred on one of my forefathers by the King of Ladakh during the eighteenth century. He distinguished himself against raiders from Turkistan (Sinkiang).

I was born on 11 November, 1931. This coincided with the auspicious 15th day of the first month of Ladakhi/ Tibetan calendar in the Year of the Horse. It is supposed to be the day when Lord Buddha won the hearts of all his opponents and spread Buddhism in whole of India.

My village, Sumur, in the Nubra Valley is located at a height of 10,000 feet in the shadow of the mighty

Karakoram range, the second highest in the world. A few months after I was born census was conducted in our village. The Patwari (the ground level revenue official) had to enter a name for me in his record and he did it arbitrarily. I was shown as Chhewang Rinchen in the census that was compiled. As per the Ladakhi tradition, a child has to be named by the Kushak, the holy head priest of the region. So my parents took me to Lama Padma for this purpose. He performed the necessary rituals and chanted the scriptures and finally sat in contemplation to choose a name for me. He did not know that the village Patwari had already recorded a name for me in the official documents. So it appeared providential when the Lama also pronounced my name as Chhewang Rinchen. Chhewang stands for a hero and Rinchen means full of life !

Overlooking our village is a hill. On top of it is the famous Gompa named Samstanlink. It is the major monastery in the Nubra Valley. Hundreds of Lamas live there for spiritual education and to preach Buddhism. They also train young children who are offered to the Gompa at a young age. As per the custom in our region parents have to offer a son or a daughter to the Gompa, at about 6 to 8 years of age, to become a celibate Lama or a Nun for life. My father was also thus offered to become a Lama. However, he gave up the Gompa life and got married. It was not easy to move out from being a Lama because it involves stringent procedures of apologies and heavy fines. Very few people ever succeed. I came to know about the episode of my father being an ex-Lama in a funny manner. Whenever, as a child, I quarrelled with our neighbour's children they would abuse me by calling me a 'Banthugu', which means the son of a Lama !

My father, Shri Kunzang, was highly qualified in Ladakhi as well as Tibetan languages. Because of this and the respect he enjoys in the region, he was requested by Mr. Kalon Rigzin, the Prime Minister to the Raja of Ladakh, to be the General Manager to look after his vast lands and properties in the Nubra Valley.

My mother Jamyang Dolma was a very nice and a hard-working woman. For her loving and kind nature she is known as the mother of Nubra Valley. She is fond of Chhang (the local beer) and is very hospitable. During one of the operations when I led a raid on the enemy post at Hundiri (opposite the present Thoise airfield), we killed many Pakistani invaders by throwing grenades in their trenches and dug outs. When my mother came to know about it she sent a special message to me not to kill the enemy but take them prisoners. Such is her compassionate heart.

I was sent for primary education to Mr. Stanzin, a Ladakhi Christian missionary who was living in our village during the British regime. He was known as the Reverend Master. Our first lesson was to pay respect to the elders and greet others with the Ladakhi word *Jullay* (meaning Jai Leh or glory to Leh). Whenever I went to their house I would greet them before opening the door "Jullay Bale Male" (meaning greetings to Abale - father and Amale - mother). One day I did this but when I opened the door only Amale was there and this Bale Male became a big joke. The reason was that as a child I had a terrible stammer and had difficulty in saying full Abale and Amale. My parents and brothers used to tease me by calling me 'Khadik' meaning a stammerer. I have now grown out of it to a large extent. We were four children. After me were two brothers and then a sister.

As a child it was my hobby to make pistols, guns and bombs. Muzzle loading guns were available and I used to make my own gun powder. I enjoyed playing with improvised weapons.

When I finished my primary education, it so happened that Mr. Kalon had come on a visit to Nubra. My father must have discussed the problem of my further education with him. The result was that I was sent to Leh with Mr. Kalon for further schooling. I stayed with the Kalon family for four years and attended school. I was treated by them like their own son and like a brother by their children. Everyone was extremely kind to me.

I used to be the monitor of my class and also the band master of the school band. I was good at arts also and won a consolation prize for a painting. All this made me a popular figure in the school.

State Force Army officers often used to visit Mr. Kalon's house for a drink or a meal. From them we used to hear stories about World Wars I and II and accounts of bravery. I also used to avail every opportunity to go and see the drill and guard mounting by the State Force troops assigned to protect the Leh treasury. Whenever these troops went to fire on the range some of us invariably reached there to watch. It was thrilling to see the automatic weapons spitting out a chain of bullets.

During 1945-46 Mr. and Mrs. Kalon and my father led a Ladakhi-Tibetan trade mission called Jung Tsong to Lhasa. In a long caravan they carried with them items imported from Sinkiang (China) like carpets, namdas, silk and marijuana as also the produce of Ladakh like dry apricots, flour, sattoo etc. They returned after ten months bringing with them items like tea, salt, Chinese crockery, cloth for men and women, dresses, good quality carpets, pearls and turquoise.

VOLUNTEER TO DEFEND MOTHERLAND

In March, 1948, I passed my 8th class examination to enter the high school. We were about to begin our studies when the war clouds started gathering over Ladakh. Our motherland was threatened by Pakistan. At this stage Lieutenant Colonel Prithi Chand (his Ladakhi name was Sonam Dawa meaning Lucky Moon) of the Indian Army arrived at Leh with a detachment of the Indian Army. They had come to organise the National Guard to meet the Pakistani challenge.

I was the first one to volunteer to join the Guard. I was then 17 years old. I know that the safety of my land was more important than my studies. It was April, 1947. After a period of ten days training I was sent with Subedar Bhim Chand (Ladakhi name Tsering Spalgias meaning Long Life Fighter) to raise a local force in the Nubra Valley known as Nubra Guard.

Within a period of one month a company strength was raised and trained. They were immediately deployed in La Chhuruk and Chhangmar area (between present Thoise airfield and Turtuk).

By this time Pakistani invaders had already captured Baltistan (except Skardu), Biagdangdo and were advancing along the Shyok river. They were preparing to attack the Chhangmar position where we had reinforced a platoon of Jammu and Kashmir Infantry sent by Lieutenant Colonel Prithi Chand. Our task was also to protect the northern bank of Shyok river.

I, with a platoon of 28 youths of Nubra Guard was given the task of raiding the La Chhuruk post of the enemy on the right bank, opposite Biagdangdo. We had a ten days march to reach the enemy position. We crossed the river upstream and then crossed the two steep and difficult mountain peaks namely Utmaru and Warris. We totally surprised the enemy and killed all the occupants, captured their arms and ammunition and occupied the post.

Next day Subedar Bhim Chand, our gallant company commander, detailed me with a platoon known as the

Diskit platoon to go out on a patrol. We had moved only a short distance when we saw a number of Pakistani unarmed soldiers carrying some bushes. On seeing us they dropped those bushes and ran. I ordered my men to chase them in order to take them prisoner.

When my men in hot pursuit turned round a knoll the enemy opened up with automatic fire. In the lead group which came under fire was Tashi, Zangskar, Chhultin and myself. The machine gun firing at us was hardly about 50 yards away. We took cover behind a big boulder. The enemy started firing at the boulder which slowly started to disintergrate. The bullets were whizzing past us. Luck was with me or the enemy machine gunner was not such a good shot; a bullet singed my woollen cap but my head was safe.

Every little moment was precious because the boulder behind which we had taken cover would break up any time and expose us to direct enemy fire. So I decided to withdraw. To draw enemy fire I rushed out first so that others could exploit the shift in enemy's attention. But that was not to be because in the spurt by others I lost a brave man Chhultin - a fine soldier from Diskit. I couldn't help but recall the deep faith of our people in God. When we were huddled behind the boulder, my friend Tashi took out a one rupee note from his pocket (a lordly sum those days) and threw it as an offering to God to protect us. It displayed the attitude and utter faith of our people in the Almighty.

After the Pakistanis captured Zojila and Kargil in May 48 the way to Ladakh was wide open. Fortunately that great fighter Brigadier Sher Jung Thapa held on to Skardu till 14 August which forced the enemy to bypass that town as it lay astride the main route to Kargil and Leh. This heroic stand slowed down his build up. But by the end of June the enemy thrust on Ladakh gathered momentum despite some reinforcements which reached Leh by air and land route (via Manali). The enemy was knocking on the gates of Leh - his troops were hardly ten miles from the town on the Indus river route. An artillery gun was being brought up to bring the Leh airfield under fire.

A CRISIS

In view of the grave threat to Leh the overall commander in Ladakh, Lieut Colonel Prithi Chand ordered all troops to rush to Leh. Indian Army elements and a platoon of Jammu and Kashmir Army in the Nubra Valley were, as a result, recalled to Leh. Subedar Bhim Chand, our gallant commander, also went away. They also took away the arms and ammunition issued to the Nubra Guards. This decision left Nubra Valley totally unprotected and completely at the mercy of Pakistani invaders. This created great panic in the area resulting in people leaving their homes and hearths and moving to find safety in Leh.

I was a witness to this mass exodus and it became unbearable for me to witness our land being offered to the enemy without resistance, virtually as a gift. I placed my agony before Lama Sondus and Maulvi Ghato Ali of Partapur village, Shri Rigzin Namgyal, Lompo of Hunder village and Shri Stanzin, the Headman of village Tagger, for their advice and guidance.

They advised that a force should be raised to protect the Nubra Valley and that they would render all assistance for this mission. Fortified by this assurance and with a burning desire to fight for the protection of my country, I went to the worship room in our home. I bowed my head before Lord Buddha, Gods and Goddesses and asked for their blessings, protection and strength to fulfil the task I had undertaken. Accompanied by Stanzin, the village Headman, I crossed over Khardungla to see Lieutenant Colonel Prithi Chand.

When we reached Leh many Nubra people agreed to accompany us for the meeting. Chhewang Rinchen, the younger brother of Mr. Kalon was the first one. He had just reached Leh bag and baggage from Nubra to run away to Tibet. Mr. Kalon himself also joined us on our ardent requests. We met Lieutenant Colonel Prithi Chand at the Zorawar fort and requested for 100 rifles and some LMGs to defend Nubra. He first hesitated to release these weapons to a mere 17-year old boy. At this stage Mr. Kalon guaranteed that no arms and ammunition will be allowed to fall in the hands of the enemy. On this assurance he finally released 28 rifles, one sten gun and a few boxes of ammunition. He also got us a few ponies to carry these to Nubra.

The next day we crossed over the Khardungla pass. We were having a short break when I gave initial training on rifle to Stanzin and made him fire a few rounds on a target. He later enlisted in the Nubra Guard as my second-in-command. As we moved on we met a caravan of people on their way to Leh. We impressed on them the vital necessity of their presence in Nubra to help our determination to protect their homes from the invaders. Our pleas had an effect and most of them agreed to return to their homes.

THE DEFENCE OF SKURU, THE GATEWAY OF NUBRA

On reaching Nubra Valley, with the assistance of the village headman of Hunder all the Nubra Guards were recalled to duty and armed to the extent possible - many with muzzle loading guns. With this ill-equipped force I left for Skuru.

Skuru was a deep and fast flowing nullah. In summer it becomes a torrent and a formidable obstacle. When we reached the village Skampuk we were given the information that Pakistani invaders had become aware of the fact that the Indian Army had pulled out of Nubra and there was no resistance. In fact they had sent a demand to village Partapur asking for 200 riding ponies, 100 draught ponies and 300 porters to carry their force to Leh via Khardungla pass. I also learnt that they would start crossing the Shyok river in local boats near the embankment of Skampuk village. In this area the boats are actually rafts made of inflated animal hides. On hearing about this plan I immediately decided to ambush them while they were crossing the river.

At about ten O'clock in the morning the Pakistani troops reached the opposite bank. A section of about 12 men boarded the boat and set out to cross the Shyok. It was right opposite to the place where we were lying in ambush. When the boat reached midstream we opened fire and all of them were killed or drowned. The enemy on the opposite bank returned with fire but were totally surprised by our resistance. After a while they withdrew towards Hundiri village on the northern bank.

Leaving a few men to guard the crossing point I moved rapidly to Skuru nullah to organise a strong defence. I evacuated the civilian population of Skuru village to Thoise after enrolling all the young men to join us with whatever weapons they had. I also destroyed the Skuru nullah bridge to ensure that the enemy could not secure it for crossing.

Next day the reinforcements arrived from the Nubra Valley. People armed with swords, spears and muzzle loading guns joined my force. They also brought provisions like sattoo, flour, meat, apricots and chhang (local beer). The driving force for all this was my father Kunzong, Rigzin Namgyal Lompo and Chhewang Rinchen Kalon. The force swelled upto 300 men - the whole of Nubra Valley had risen to a man. The Pakistanis misjudged this force as being the Indian Army. The result was that they desisted attacking our position for a few days. Headman Stanzin who was my second-in-command with three Nubra Guards and 30 volunteers was sent to Hunder via Kubet bridge to keep a watch on the enemy's activities and to guard that area.

After about eight days of waiting and preparation the enemy attacked us with a company strength in the early hours of morning. The attack was supported by machine gun and mortar fire, but it was repulsed with heavy casualties to the enemy while we suffered only one casualty. After this set-back, the Pakistanis occupied a defensive position on the opposite bank of the Skuru nullah and indulged in intermittent firing on our position. The headquarter of our position was located in a Gompa overlooking the nullah. The Pakistanis used 2-inch mortar fire to destory the Gompa but fortunately most of the bombs proved to be dud.

Deception played a big role in our strategy. To mislead the enemy about our strength and fire power I carried out raids on them from different points and directions. We also used to burn fires at numerous places to depict a large strength. Loaded ponies were shown approaching some positions in the evening, withdrawn during the night with their loads and shown approaching a different area the next day. All this depicted an extended front and great depth to our defences. After a few days, the enemy launched another attack at about mid-night using hand grenades. This resulted in a few casualties on our side but our men held their position and the attack was beaten back. In this battle two men showed outstanding valour - Tsering Wangdas (he was later mustered into 7 Jammu and Kashmir Militia and rose to be an officer) and Chewang Targais (he rose to become the headman of village Hunder).

By now it was August. While whole of the 2/8 Gorkha Rifles and Lieutenant Colonel Parab (he was given the rank of Colonel and appointed Governor of Ladakh) had reached Leh. The enemy reinforcements were also pouring in. In village Baigdangdo lived a versatile artisan who was expert in many mechanical trades, named Mistry Hassan. He was a friend of my father and also his business partner in trading Yaks, dried apricots, salt wheat, etc. He managed to send a runner to my father with a secret letter. The Pakistanis were expecting reinforcements any day and had planned to launch a major attack on Skuru defences. They specially wanted to capture or kill Chhewang Rinchen. He advised my father to alert his son (me).

On getting this information I went to consult Rigzin Namgyal Lompo. He sent a special runner to Lieutenant Colonel Prithi Chand with this information and to request for immediate reinforcement, arms and ammunition. We received an immediate reply with the following letter to me:

"My dear brave Rinchen, Commander Nubra Guards. I am in receipt of your and Shri Rigzin Namgal Lompo's letters. We understand that you and your brave men are fighting against regular Pakistan Army. You have played a major role for the defence of Nubra and Leh fronts. Our troops are fighting bravely on the Leh front and both sides have suffered heavy casualties. We are getting reinforcements by air and land route via Manali. I am sending you reinforcements, also more arms and ammunition for a company strength of Nubra Guards; till such time Lord Buddha is with you.

16 Aug., 1948 Signed
Thakur Prithi Chand

Battle was raging on the Skuru front. It was mainly sniping and shooting between the two sides. On the morning of 23 August I was surprised when my 15-year old younger brother Phunsog Namgyal (later a member of Parliament and a Central Govt. Minister for many years) entered my position just behind the Skuru Gompa. I asked him how on earth he had reached there during day time when the slightest movement attracted fire. He replied that he was accompanying Lieutenant Colonel Prithi Chand and party, but due to enemy firing they went back to Tarche village. But he being a small target, so he crawled along a dry water channel and reached the post. He conveyed a message that I should meet the Colonel at Tarche. My runner Tsering Murup of Nubra Guard quickly brewed a cup of tea for Namgyal who was very keen to see enemy defences in front of our position.

I asked Angdan, who was acting as our Quarter master, to take him and show him the defences very carefully. When they were coming back along the crawl trenches the enemy spotted them and fired a few bursts of automatic fire which hit Angdan on the head. Namgyal came running to tell me that Angdan had been hurt by enemy bullets. I rushed to see him and found that he had died on the spot. After a few minutes a runner came to inform me that Guardman Tashi was also hit by an automatic burst. After disposing of the dead bodies I went to Tarche to meet Lieutenant Colonel Prithi Chand.

When I reached the village and entered the Headman's house I found Lieutenant Colonel Prithi Chand as also my father waiting there. They welcomed me warmly and the Colonel even garlanded me and praised me a great deal. He then informed me that he had brought a platoon of newly raised 7 Jammu and Kashmir Militia along with a company of 2/8 Gorkha Rifles under command of Major R.C. Mathur, the second-in-command of the Militia battalion. He directed me to take them to Skuru defences, explain to them our and enemy disposition and report back to him at Tarche at first light to proceed to Diskit.

After holding the position for 23 days I handed over the post to Subedar Ishar Singh of 7 J & K Militia and also explained to him Mistry Hassan's message. The Gorkhas were also deployed. A few days later the enemy launched a major attack and captured the Skuru post. They also captured a section post of 2/8 Gorkha company. Our troops fell back to Skampuk village. There, due to bravery and excellent skill of Major SJS Bhonsle (second-in-command 7 J & K Militia and Sector Commander) our troops recaptured the feature ahead of the village named Skampuk top by a platoon of the Gorkhas. In this action Subedar Nar Bahadur displayed outstanding courage and was awarded a Vir Chakra. The message of Mistry Hassan proved right.

LOCAL GOVERNMENT IN NUBRA VALLEY

On 25 August, 1948 there was a grand reception in my honour at Diskit Village, the capital of Nubra valley. Lieutenant Colonel Prithi Chand addressed the people and appointed me as a Naib Subedar in 7 Jammu and Kashmir Militia (the youngest ever Junior Commissioned Officer in the Army). There was no uniform for me, so a set was taken from a Naib Subedar of the Gorkhas and I was ceremonially given the badges of rank.

The Colonel had also to appoint a local government in Nubra Valley to work under Colonel Parab, the Governor of Ladakh. The following were appointed :

Shri Chetan Phunsog	-	Administrative member
Shri Rigzin Namgyal	-	Defence member
Shri Chhewang Rinchen Kalon Lompo	-	Supply member
Shri Kunzong Dorje	-	Political member
Shri Stanzin Dorje Kaga	-	Transport member

They worked effectively during the rest of the period. I was assigned the task to once again raise a full company of Nubra Guard and select 50 men to operate as guerrillas. These specially selected men were given training in the use of Light Machine Guns, 2-inch mortar and all types of grenades. I was assisted by instructors from the Gorkhas for this training. On completion of this training I was aksed to take over command of this group and to undertake a number of operations.

BATTLE OF THE LAMA HOUSE

The first task given to me was to capture the Lama House which was under occupation of Pakistani troops near village Hundiri (opposite the present Thoise airfield). It was early September but snow had already fallen on the higher mountains. I proceeded on my mission and established a base in Nullah Chakri Chubab. Next morning we started on manpack basis for attacking the Lama House. Our ponies were left in the base.

Our guide on this mission was Zamkhan Chhewang, a resident of nearby Charusa village who knew the area well. During this march Zamkhan and I led the way. That night we spent at 15,000 feet. Next morning we reached the top of the mountain at 17,500 feet at about 11.30 in the forenoon. From here we could see the Pakistani post at about 15,000 feet on the opposite slope. Therefore our attack involved first going down and then climbing the opposite slope. After lunch, Havildar Tashi Mutup, guide Zamkhan Chhewang and myself left to reconnoitre the route and the Pakistani post.

We discarded our red and long Gonchas - the normal Ladakhi dress - and crawled down the slope like Ibex. We reached about three hundred yards short of the enemy post and carefully observed the layout of the defences. I estimated that the post was manned by about a platoon as we counted 25 men who were engaged in constructing a perimeter wall. Their living accommodation was at the centre of the post.

On returning to our own hill top I briefed my men in detail. As one of my soldiers had fallen ill, the guide Zamkhan offered to take his place. Carrying a rifle he closely followed me through out the attack.

I divided my men into three groups. One group to approach the enemy post from the right flank, the second group in the centre and the third group under me to launch the attack from the left. We started soon after it was dark. Since the slope was steep and covered with snow the descent was very difficult and slow. At times we had to slide down with the help of ropes tied to rocks. We reached near the enemy post at about 3 O' clock in the morning.

After the groups had taken position as planned, I along with the third group went in for the assault. We had reached near the enemy living bunker when their sentry sighted us and opened fire. However, before the post could react, I threw a hand grenade in the main bunker and bayonet charged the post. While we were rushing forward I suddenly heard a burst of Light Machine Gun fire. On watching carefully the direction from which the fire was coming I realised that our own central group was firing on us. I shouted and ordered them to stop firing. The response from the other end was to ask me for my identity. I disclosed my identity but they continued to fire till I shouted the password for that night 'LARGIL'. They stopped firing but did what Pakistanis could not do. By this mistake they had seriously wounded Lance Naik Nurboo.

Major portion of the enemy platoon were killed but many ran away in their under clothes. Among the dead was their platoon commander known as Sergeant Major Mota Hassan (fat Hassan) of Gilgit Scouts. Appropriate to his name he weighed about 250 pounds. I killed him in hand-to-hand fighting with a bayonet and captured his Stengun. This sten was presented to the Hall of Fame at Leh and lies there. When we reached their headquarters in the Lama House we found it deserted. The elements there had run away without offering any resistance. After a few days my position was reinforced by a platoon of 2/4 Gorkha Rifles with a section of 3-inch mortars under Subedar Moti Lal, Military Cross, and the remaining portion of Nubra Guard.

FALL OF SKURU

The day after we captured the Lama House we got the news that Skuru position had fallen. I have related this earlier also. However Major Bhonsle took effective steps and the enemy could not make any further progress. At this stage Major Bhonsle was called back to Leh and Major J.S. Sidhu of 8 Gorkhas was appointed the Sector Commander in Nubra Valley. To assess the situation in Nubra Valley, Colonel Parab who had taken over as the Governor of Ladakh visited Nubra. He ordered an attack on Pakistani position at Tarche but the operation did not succeed. The next day Colonel Parab sent for me from the Lama House.

I presented myself in my local clothes. I was wearing a long goncha, a worsted woollen cap and Ladakhi long boots. I had a stengun under my arm and a few grenades tied to my belt. Colonel Parab was in the garden of Rigzin Namgyal Lompo, where Major Sidhu and Captain Engineer of 8 Gorkhas were with him. As soon as I arrived Colonel Parab pointed his stick towards me and enquired, "Is this the boy who has been fighting the Pakistanis?", and then went on to praise my bravery. The meeting ended by his ordering me to capture and bring back the Pakistani Medium Machine Gun which was firing at our recently captured post in the Lama House.

That night I along with three volunteers went for this mission. We carried our stens and a few grenades each.

Our plan was to crawl close to the MMG position, throw grenades into it and then lift the Gun after killing the crew. All went as per the plan. We were hardly 30 yards away from the gun when a stone on the slope got dislodged and rolled down. On this the sentry got alerted and raised an alarm by shouting 'Dushman, Dushman' (enemy, enemy). In this confusion, I threw a hand grenade in the enemy post and then we ran for cover. We had hardly reached safety when the enemy opened up with the MMG and also threw a few hand grenades. We made our way back and reported to the Sector Commander about the failure of our mission.

TIFF WITH MAJOR SIDHU

I had selected a very fine riding pony for my use on duty. One day I received a message from Major Sidhu ordering me to send that pony to Sector Headquarters at Hunder for his use. I flatly refused. He sent for me to report to him at the Sector Headquarters. When I went into his office he appeared annoyed and used harsh words to the effect, "You think this is your private horse. I shall have you dismissed from service." I replied in the same manner, "I want to be dismissed," threw the sten gun on his table and walked away to my house.

The very next day I was summoned to report to the Governor at Leh. It was a personal message from Colonel Parab. I reached Leh but before going to the Governor, I went and met my Commanding Officer, Lieutenant Colonel Prithi Chand. I reported the incident exactly as it had happened. He listened patiently and stated that I was indispensable for the defence of Ladakh. So he ordered me to take over a platoon of 7 J and K Militia (all Ladakhis) and go to the Leh front. He also expressed surprise at my civilian clothes and ordered the Quartermaster to equip me with proper uniform.

Something must have then transpired between Lieutenant Colonel Prithi Chand and Colonel Parab. I know only the result. I was ordered by the Governor to go back to the Nubra Valley and take over the command of the entire Nubra Guard and the Guerrillas. With these orders I presented myself to Major Sidhu. He in the best tradition of the Army and as a fine soldier shook me by the hand and invited to have a drink with him to celebrate my return. All was over.

It was getting on to the end of November now and the happy news of General Thimmayya's success in capturing Zojila reached us. We also learnt that the Indian Army was rapidly advancing to reach Kargil. We also got information that the enemy opposite us was preparing to pull out. Major Sidhu ordered his force including the Nubra Guards to be ready for advance.

OUR ADVANCE

We advanced methodically on both banks of Shyok River and occupied Chhangmar La, Chhuruk and Baigdangdo.

Before we reached Baigdangdo, the headman Mistry Hassan carrying a white flag came to meet us. He assured us of the village's support and sympathy for our Army. When we entered the village our forces were welcomed by the local population. However, there were no women and children. We were told that they had run away and were hiding in the nullah in the hills. They had done so because they had suffered atrocities at the hands of the Pakistani invaders. They feared that the same would be repeated by the Indian Army.

I was detailed by Major Sidhu, the Sector Commander, to convey to the women our assurance of good behaviour and persuade them to return to the village. Accompanied by a section of Gorkhas and some villagers I went to the nullah. When they saw us they started to climb the adjoining hills. With great difficulty we were able to stop them and gather them in one place. I spoke to them in their own dialect and explained to them that the soldiers of the Indian Army would treat them as their own mothers and sisters and have come to protect them from the Pakistanis. I told them that they should have no fear in their minds at all. Finally they were persuaded to return to their villages.

Our advance was halted at Baigdangdo as Pakistanis had occupied a strong position in the Black Rock area. They had also occupied outposts on Takkar hill next to it. Orders to attack these were issued. On 15 December the main force of Gorkhas captured the Black Rock position and I captured the Takkar Hill with the Nubra Guard troops. We inflicted heavy casualties on the enemy and they fell back to Chalunka.

The enemy was still holding positions in the Tebe Hill area on the South of the river. This position dominated the whole valley and our advance was held up. On 22 December I was ordered to capture Tebe Hill position and to reach Tebe Nullah. My Guerrilla platoon and twenty young men of the village Biagdangdo as porters were organised for this task. We crossed over the Shyok river. It was snowing and by the evening we reached the base of the high mountains and spent the night there.

Next morning, I with a small group of volunteers started climbing the mountain. The rest of the force followed our footsteps as we selected and cleared the best possible route. Stanzin Tundup ably assisted me for this work. We didn't want to use the normal route and were moving on a much higher contour. At sun set we reached the mountain top at 21,000 feet and waited for the rest to join us. But there was no sign of them. We shared the home cooked thick bread, some sattoo and a bottle of local liquor. We dug the snow and slept like the Ibex. Inspite of the extreme cold and very high altitude none of us suffered from frost-bite. The next morning we tried to observe the Pakistani position but could not see it from the height where we were.

At about 9 O'clock, one section under Rigzin Phunchok joined us. They reported that they had to leave all the beddings and rations behind because a number of porters and soldiers were suffering from frost-bite. The remaining force joined us at about 11 O'clock. After eating some *sattoo* and tea we started our descent. It was treacherous and steep going. By the evening we reached a shoulder on the side of the mountain covered with snow which was about 19000 ft high and was just about 1000 ft above the enemy post. It was a very useful position as we could see them but they could not see us. So I decided to spend the night on this shoulder and work out our plan. We had dinner of hot *thupa* (noodles made out of flour). We also discussed our options to evolve the best possible plan. We decided that I with my platoon headquarter and one section will lead the assault. One section under Chering Spalgais would give fire support and the third section under Rigzin Phunchok was to swing left and cut off the enemy on the ridge where his main route to the post lay.

Next morning we started early. When we were about a 1000 yards short of the post we found that a deep nullah separated us. It was decided that the section under Spalgais would take position while I crossed the nullah. He could give me fire support in case we ran into the enemy while climbing after the crossing. Fortunately we climbed the far side of the nullah without being noticed. When we reached the top we were hardly 300 yards from the post and the day was clear with the sun shining in the deep blue sky. I carefully observed the Pakistani post. About a section strength was on duty in the post. While 25 men were outside basking in the sun and smoking a *hukka* (hubble-bubble). At this stage I was carrying the 2-inch mortar and porter Mohamad from Baigdangdo was carrying the bombs. I ordered him to take out the bombs from the container.

Targais Hunder who was carrying the Light Machine Gun was also with me. I took his LMG and fired a long burst with one full magazine at the hukka smoke party. About six or seven were killed and the remaining ran into bunkers. I then shifted to the 2-inch mortar and bombarded the post. Fortunately one bomb scored a direct hit on a big bunker. The result was that the enemy started running. Simultaneously our 4.2-inch and 3-inch mortars from the northern bank of the river in Black Rock area was also brought down on the enemy. Rigzin Phunchok on the left flank also got into a good position and started engaging the enemy.

After the enemy was fully pounded I ordered a bayonet charge. On entering the post we captured some wounded soldiers. A search of the dead bodies was also carried out and we found large amount of cash and turquoise which they had looted from the local population during their occupation of the area. While I was mopping up the enemy post an unfortunate incident took place. We suddenly came under heavy fire. The very first burst by an LMG killed porter Mohamad who was following me like my shadow and I providentially escaped - but barely. We immediately took cover and tried to locate the position from where fire was coming. I realised that it was from our own covering section. Obviously they had mistaken us to be the enemy. I shouted to the section but could not be heard. I then ordered a soldier to take off his woollen monkey cap, which we were wearing, put it on his bayonet and raise it from behind a banker. He did so. This signal proved effective and the firing stopped.

We soon set out in pursuit of the enemy. On the way we came across a few wounded and many dead Pakistanis. One of the wounded fellows whom we took prisoner informed us that the Pakistani company headquarter was located in the Tebe Nullah very near the track. We pushed on and reached Tebe Nullah at about midnight. We found the company headquarter but it was deserted. Having reached our final objective I conveyed our success to the Black Rock position by firing a very light pistol round. It had been a long, exhausting and hazardous operation lasting for six days, of which four nights and days were in contact. We felt we needed a celebration and had a good hot meal cooked from the rations left behind by the fleeing Pakistanis.

THE END OF THE BATTLE

On 1 January we were taken by surprise to hear the news of a ceasefire. We felt we were on our way to recover Baltistan from the enemy but such are the events of history. Once the ceasefire stabilised the Pakistani sector commander invited our sector commander for a lunch at their headquarter at Chulungkha village. Our sector commander accepted the invitation and took me and a few soldiers along. We were received most cordially, the local population staged a cultural show in our honour after having turned out to receive us.

It was a very friendly occasion and when we were leaving the Pakistani sector commander presented each one of us with insignia of Gilgit Scouts and a pair of woollen socks. After the ceasefire was fully stable I went on leave-cum-recruitment drive for Jammu and Kashmir Militia. When I reached Nubra Valley I was given a hero's welcome and ballads were sung praising my actions.

On 26 January 1950 the first Republic Day was celebrated by India. On that day during the parade and celebration the citation for my Maha Vir Chakra was read out and the medal awarded to me by the President. It was a happy and a proud day in my life.

In 1952, when Jammu and Kashmir constitution was implemented, a similar investiture ceremony was held at Srinagar in September 1952. Sheikh Mohammad Abdullah who was the Prime Minister of the State awarded me the honour. Later, another public function was held at Leh when a silver shield was presented to me on behalf of the people of Ladakh.

Epilogue

In 1947 the British Rule in India ended with the partition of the subcontinent. It was the logical culmination of the ancient and time-tested imperial policy of governance - divide and rule. It is our own fault that we became a prey of this policy.

The madness that gripped Hindus, Muslims and Sikhs at the time of the partition resulted in the great tragedy of mass killings and millions of people being uprooted from their homes and hearths. What happened in Jammu and Kashmir was a by-product of this tragedy; the Ladakh story in this book was a part of it.

Fifty years have gone by, but there are no signs of reconciliation between India and Pakistan. The author and his wife have worked for thirteen years for this cause and have come to the conclusion that people in the two countries are yearning for it. Only the *establishments* in India and Pakistan stand in the way. It would be appropriate to record the TRUTH of the major events in this tragedy because TRUTH ultimately prevails :

> "By Time, verily man is in loss, except such as have faith, and do righteous deeds, and in mutual teaching of *Haq* (TRUTH) and patience and constancy"
>
> - Sura 103

"Sathyam - eva - Jayathe" - TRUTH always truimphs.

- Upanishads

The TRUTH about the nine major events, during the half century that has rolled by, which can help people of the two countries to view their misguided hostility with clinical detachment are :

ONE : India was partitioned to provide a homeland to the Muslims of India. This, unfortunately, has not happened despite a very heavy price paid in human suffering in the biggest migration of history. It is time that people of India and Pakistan took a dispassionate look at their hurts and hates to see what is in the best interests of the people of the two countries. In the event, partition has turned out to be a great tragedy for the Muslim community of undivided India; it has been fractured into three nation states :

- The largest segment continues to be in India where it has shouldered a sense of guilt, has been suspected, exploited as a vote bank and in some areas has withdrawn into a ghetto mentality.
- The second segment is in Bangladesh which is the most thickly populated country in the world with very few resources.
- The third segment is in Pakistan where it has been directly or indirectly ruled by the military since 1958. Those Urdu speaking migrants who went to the 'Muslim homeland' are either languishing in refugee camps in Bangladesh or having a rough time in Sindh.

A portion of the Pakistan establishment refuses to see this as the stark TRUTH. Similarly in India the establishment dismisses this TRUTH with a brusque remark - "well, the Muslims asked for it." These attitudes are neither wise nor humane. Partition of India indeed was a monumental tragedy. It cannot and must not now be undone. But reconciliation can mitigate its ill-effects.

TWO : J and K problem is merely a symptom of the tragedy of the partition. It was precipitated when Liaquat Ali as the Prime Minister and Defence Minister assisted by Iskander Mirza, the Defence Secretary (later President of Pakistan), Khan Abdul Qayum Khan, the Chief Minister of NWFP, and Ishaq Ahmad Khan working for the Chief Minister of NWFP as a provincial civil servant (later President of Pakistan) decided to annex J and K by force, bypassing the Army hierarchy. Colonel Mohamad Akbar Khan[1], DSO, the Director of Weapons and Equipment in the GHQ was used to evolve a two - phase plan (Akbar[1] later adopted the romantic name of General Tariq while leading the Tribal Invasion). In Phase-one 4000 rifles were issued, mainly in Poonch area, to trigger a rovolt by the large number of demobilised Muslim soldiers who had served in the Indian Army during World War II. In phase-two, the Tribal Invasion was launched on 20 October 1947. The Tribal Lashkars (Forces) were led by officers like Major Anwar Khan, a demobilised emergency commissioned officer of the ASC, who was the commander of the Lashkar (force) earmarked to capture Srinagar. *The only remuneration that was agreed to with the Tribal Lashkars was 'to loot the non-Muslims'*[2].

The Invasion has been described in the book *Memories of Jinnah* by his Kashmiri Secretary as *a criminal folly which sealed the fate of Kashmir*[3]. It appears that Jinnah was informed of the invasion plan only on 11/12 Oct. 1947 that it was to be on 20 Oct.[4] He was assured of success and was invited to be at Abbottabad for a truimphal drive to Srinagar. When the plan did not succeed Liaquat put forward the excuse of a conspiracy by Mountbatten and Nehru to deprive Pakistan of the State of J and K. It was obviously an alibi for the failure by the organisers of the invasion; that is the TRUTH.

THREE : When the Tribal invaders were nearing Srinagar the Maharaja of J and K offered to accede to India if the Indian Army could be sent to save Srinagar from the fate of the loot, arson and rape meted out to Baramula. Mehr Chand Mahajan, the Prime Minister of the State, flew to Delhi and persuaded a reluctant Nehru to send the Indian Army with these words, "Give Army and take accession or I will fly to Lahore to negotiate terms with Jinnah."[5] Nehru got angry and asked Mahajan to go away. But Sheikh Abdullah, who was staying with Nehru, interceded and accession was accepted. Srinagar was saved by the Indian Army troops flown in. It was a touch-and-go affair.

While accepting the accession, Mountbatten conveyed the Indian Government's position that "accession will be finalised by a reference to people after the invader is pushed out of the State". The TRUTH is that this has not happened and the invader is still in occupation of what the Indians call the 'Pakistan Occupied Kashmir' and Pakistanis call it the 'Azad Kashmir'.

FOUR : India complained to the United Nations against this aggression. That body in a resolution of 13 August 1948 and other linked to it, postulated a three-step solution:

* Step 1 — A ceasefire. This took place on 1 January, 1949.

* Step 2 — Pakistan to pull out of the State, this has not happened so far.

* Step 3 — Future of the State to be decided by a reference to the people. When step 2 was not taken, India got the accession ratified by the Constituent Assembly of J and K to complete the process of reference to the people. Pakistan integrated northern areas (Gilgit-Skardu) into Pakistan.

FIVE : In 1962 India suffered a humiliating defeat at the hands of the Chinese. The British government promoted a dialogue between India and Pakistan to settle the J and K problem. Bhutto representing Pakistan for talks with Swaran Singh of India demanded the whole of the State (98% land) leaving the Tehsil of Kathua to India. The talks failed.

SIX : Bhutto as the Foreign Minister of Pakistan bypassed the Army hierarchy and got the 'Kashmir group' to prepare 'Operation Gilbraltar' supported by 'Operation Grand Slam' to annex J and K while India was still dazed with the Chinese trauma. The plan involved infiltrating soldiers in civilian clothes across the ceasefire-line into Kashmir to trigger an uprising and severing the Indian lines-of-communication with Operation Grand Slam. When General Musa, the Pakistani Army Chief, learnt of the plan, he objected to it on two counts. First, the Pakistani intelligence agencies had not yet prepared a dependable network in the State to trigger and sustain an uprising. Second, the operation would surely escalate into an all out war for which Pakistan was not ready[6]. Musa's objections were eventually overruled by the President, Field Marshal Ayub Khan, and Operation Gibraltar* was launched in August 1965. It escalated into an all-out war. Musa was proved right on both counts and the State could not be annexed. Under the Tashkent agreement both sides agreed to revert to status-quo-ante.

SEVEN : During the late sixties, Pakistan intelligence agencies started a long-term plan to prepare a proper covert network of committed supporters in the State for an

---

*Colonel SGS Mehdi, an outstanding officer who was an instructor at the Indian Military Academy before partition warned the Army hierarchy, in writing, that Operation GIBRALTAR will end up as the 'Bay of Pigs adventure' by the American in Cuba. The reason was that the Operation was poorly planned and prepared.

uprising at an appropriate time. Islamic sentiment and an utterly corrupt system of governance in J and K were fully exploited. It was this development which prompted me to formulate the subject for my compulsory dissertation during the year-long National Defence College course of 1971; it was 'Preventing Insurgency in Jammu and Kashmir.' I had gone to attend the course after commanding a brigade in Kashmir and the fact that an insurgency was being prepared was an open secret. But Delhi refused to face the TRUTH and take action to prevent it.

EIGHT : In 1971 East Pakistan erupted. India exploited the situation and launched what is frequently called Operation Gibraltar (Indian model). This was by supporting Mukti Bahini (The liberation army formed by the Bengali elements of Pakistan military). In December 1971, a full fledged war broke out resulting in the birth of Bangladesh. The Simla agreement ended that war. The clause in this agreement dealing with the problem in J and K reads as under :

> *"In Jammu and Kashmir the line of control resulting from ceasefire on 17 December 1971 shall be respected by both sides without prejudice to the recognised position of either side. Neither side shall seek to alter it unilaterally, irrespective of mutual differences and legal interpretations. Both sides undertake to refrain from threats or use of force in violation of this line."*

The portion in italics in the above article was inserted at the last minute when Bhutto undertook to convert the line of control into a permanent border by integrating Pakistan occupied Kashmir into Pakistan. The TRUTH is that he did not do so.

NINE : Manufacturing threats is a fine art mastered by most militaries of the world. Pakistan military has convinced the Pakistani *establishment* that India would capture large portions of Pakistan unless seven per cent (Plus/minus of the GDP; which is around 35 to 39 per cent of the federal budget) is allocated for Defence. The fact that India did not annex any portion of East Pakistan does not dissuade them from a conviction in this UNTRUTH. Bulk of this allocation (over 70 per cent of it) is spent on manpower related costs as Pakistan has acquired the most expensive manning system among the large countries of the world. Indians are not far behind (!!) even though for Defence they spend only 3 per cent (plus/minus) of their GDP which is nearly eight times larger than that of Pakistan. Many Indian strategists see in this situation an excellent arrangement to drive Pakistan to bankruptcy like the USSR which broke up by over-spending on Defence to match the USA which had five times larger GDP. These Indians see the proxy war launched by Pakistan in Kashmir since 1989 something with which India can live and argue for paying back Pakistan in the same coin. This approach, unfortunately, will only retard the development of this region as a whole.

With the above history of 50 years, the two countries are locked in an automatic spiral of action - reaction and the tempo of covert violence and terrorism is gathering momentum. The stand of the two countries on Jammu and Kashmir which has become the symbol of this hostility is :

India : Jammu and Kashmir is an integral part of India. The only unfinished business is to persuade Pakistan to vacate Pakistan-occupied Kashmir.

Pakistan : Majority of the people of J and K are Muslims.They should have the right of self-determination by a plebiscite to choose between India and Pakistan.

After eleven years of efforts for India - Pakistan reconciliation, we have come to the conclusion that the *establishments* in the two countries cannot end hostility. The fate of Nehru - Liaquat agreement and Simla agreement also suggest that even signed agreements are worth nothing in any case.

The only way attitudes can change is by people-to-people contact with a focus on the TRUTH of our problems. Absence of open, sincere and truthful communication is the cause of all fractured relations in families, institutions as well as in and among nations. Fortunately there is a strong undercurrent of love and affection among people of the two countries. Perhaps it is due to centuries old deep roots in the geography of this whole region. Even among the *establishments* of the two countries there are large, though silent, sections who realise that hostility is essentially promoted by those who have a vested interest in perpetuating it - intelligence services, traders who handle business between the two countries via third countries and make huge profits, certain sections of clergies and, of course, a category of the military elite in the two countries.

India and Pakistan will do well to take a note of the irreversible mega trend in the evolution of mankind towards *unity of man, global economy and earth citizenship.* This trend has been summarised in the following words by an eminent scholar[7] :

"The world capitalism is in a crisis. Crisis might take 25 to 50 years to play itself out and at the end of it we will see some new system created. The days of nationalism of hope (leading to the end of Colonial era) and nationalism of despair (noticeable in the experience of the USSR and ethnic-religion based national liberation movements) are also over."

This surge of humanity towards a new age is supported by another eminent American playing a daily role to move the world towards a global economy. Robert Bartley, the editor of *Wall Street Journal* in a seminal essay on America's future role opines that it should play a leadership role to "build international institutions appropriate to the world order waiting-to-be-born and to manage the friction it will create. It surely does not mean creating a new United Nations or other world-government-at-a-stroke. But it may well mean a broader application of the slow, step-by-step, economic first evolution towards unity we are already witnessing in the European Community."[8]

What is the new age waiting-to-be-born? The blue-print for this era has been articulated in unambiguous terms by a man[9] of infinite wisdom in India :

"There is only one nation, the nation of humanity;

There is only one religion, the religion of love ;

There is only one language, the language of the heart ;

There is only one God, and he is omnipresent."

South Asia is destined to play an important role alongwith North America, Western Europe, Russia, China and Japan to move mankind to the new age. The coming generations is India and Pakistan will never excuse us if we do not learn a lesson from Western Europe and bury our hurts and hates. Let us be worthy of our children and grand children.

NOTES

1. Mohamad Akbar Khan, Major General, DSO *Raiders in Kashmir*; Pakistan Publishers, Karachi, 1970, and his interview with Brigadier A.R. Siddiqi published in Defence Journal (Karachi, June-July 1985).

2. Ibid.

3. Khurshid K.H., *Memories of Jinnah,* Oxford University Press, Karachi, 1990. Also audio recording of the discussion on BBC on 25 Nov., 1991, in the series 'Conflicting opinion' on Jammu and Kashmir. Participants were : Professor Alastair Lamb, author of the book *Kashmir: Disputed Legacy,* Sir Mark Tully, the BBC representative in Delhi for 20 years, and Dr. Wiqar Ahmad of Urdu Services of the BBC.

4. There is an interesting entry in the Diary of Sir George Cunningham, the Governor of NWFP on 26 Oct., 1947: "Apparently Jinnah himself heard of what was going on about 15 days ago (i.e. around 11/12 Oct., as the invasion was launched on 20 Oct., 1947), but said, "Don't tell me any thing about it. My conscience must be clear." See quote by A.G. Noorani in his review of the book *'Kashmir : Disputed Legacy'* by Professor Lamb in *Frontline* magazine of 27 March, 1992.

5. Mahajan Mehr Chand, Chief Justice of India, *Looking Back-an Autobiography,* Asia Publishing House, New Delhi, 1963.

6. Mehdi S.G.S. Colonel, Mehdi Papers Serial three, Mehdi Foundation, Karachi, 1987.

7. Prof. Immanuel Wallerstein, Director of Fernand Braudel Centre for the study of economics, historical systems and civilisations at the State University of New York, Binghamton and President of World Sociological Congress in conversation with Mohamad Asif published in the *Sunday Times of India* dated 6 April, 1997.

8. *The New American Vision* (Macmillan New Delhi, 1991) is a collection of sixteen essays by eminent thinkers on America's role in the world after winning the Cold War. The quoted view is of Robert Bartley,

who as the editor of Wall Street Journal, is seeing virtually the daily progress in the integration of the global economy.

9. Sathya Sai Baba who spent a life time in a mosque is now operating from a global centre to promote human values of truth, love, right conduct and equanimity and is the Chancellor of Sathya Sai University at Puttaparthi near Bangalore.

Bibliography

Akbar Khan, DSO, Major General — *Raider in Kashmir,* Pakistan Publishers, Karachi, 1970.

Bamzai, PNK — *Kashmir and Power Politics : From Lake Success to Tashkent,* Metropolitan Books, Delhi, 1966.

Bedi BPL and BediFreda — *Sheikh Abdullah : His Life and Ideals,* Srinagar, 1949.

Birdwood Lord — *Two Nations and Kashmir,* Roberthale Limited, London, 1956.

Campbell-Johnson, Allan — *Mission with Mountbatten,* Roberthale Limited, London, 1951.

Chibber ML, Lt. Gen. Dr. — *Military Leadership to Prevent Military Coup,* Lance Interna- tional, New Delhi, 1986.

History of Jammu and Kashmir Light Infantry, J&K LI, Regimental Centre, Srinagar, 1992.

Chapter on 'Quest for Security', *The Indira Gandhi Years,* edited by A.K. Damodaran and U.S. Bajpai, New Delhi, 1990.

Chapter on India-Pakistan Reconciliation in *South Asia After the Cold War-International Perspectives* edited by Stephen Cohen and Bajpai, Westview, USA, 1993.

Collins, Larry and Dominique Lapierre	*Mountbatten and Independent India, 16 August 1947 to 18 June 1948,* Vikas Publishing House, New Delhi, 1984.
Das Durga	*Sardar Patel's Correspondence 1945 - 50, Vol. I, New Light on Kashmir,* Navjivan Publishing House, Ahmedabad, 1971.
Ghazanawi, Khalid	*Story of Indian Aggressions against Pakistan,* Lahore, 1966.
Government of India	*Imperial Gazetteer of India - Provincial Series - Kashmir and Jammu,* Government Press, Calcutta, 1909.
Gundevia, YD	*The Testament of Sheikh Abdulla ,* New Delhi, 1974.
Gupta, Sisir	*Kashmir : A Study in India - Pakistan Relations,* Asia Publishing House, New Delhi, 1966.
Hassnain, F.M.	*Gilgit: the Northern Gate of India,* Sterling Publishers, New Delhi, 1978.
Kak, BD	*Fall of Gilgit : The Untold Story of Indo-Pak Affairs from Jinnah to Bhutto (1947 to July 1977),* New Delhi, 1977.
Kalkat, OS	*The Far Flung Frontiers,* Allied Publishers (Pvt.) Ltd., New Delhi, 1983.
Kavic, Lorne,	*India's Quest for Security - Defence Policies 1947-65,* California, 1967, Keesing Contemporary Archives, London, 1948.

Khan, Rehmutullah, *Kashmir and the United Nations,* Delhi, 1969.

Khurshid KH, *Memories of Jinnah,* Oxford University Press, Karachi, 1990.

Krishna Rao, General *Prepare or Perish,* Lancer International, New KV, Delhi, 1991.

Lamb, Alastair *Crisis in Kashmir, 1947-66,* Routledge & Kogan Paul, London, 1961.

Kashmir : A Disputed Legacy, Oxford Book, 1990.

Mahajan, MC *Looking Back,* Autobiography, Asia Publishing House, Bombay, 1963.

Mohamad, Musa *My Version.*

Moore RJ *Making the New Commonwealth,* Oxford University Press, 1987.

Musa, Mohammad, *Jawan to General : Recollections of a Pakistani Soldier,* New Delhi, 1985.

Mushtaq Ahmed *Government and Politics in Pakistan,* Karachi, 1959.

Palit DK, Vr C, Maj. Gen *War in Himalayas,* Lancer International, New Delhi, 1991.

Prasad SN, Dr and Dharam Pal, Dr *History of Operations in Jammu and Kashmir - 1947-48,* History Division, Ministry of Defence, India, 1987.

Razvi, Majtaba, *Frontier of Pakistan : A Study of Frontier Problems in Pakistan's Foreign Policy,* Karachi, 1971.

Riza, Major General S, *Izzat-o-Iqbal (History of Pakistan Artillery)*, Nowshera, 1980.

Sen, LP, Brigadier *Slender was the Thread*, Longman , New Delhi, 1969.

Shankar, V *My Reminiscences of Sardar Patel*, Vols I & II, Macmillan Books, Delhi, 1974.

Siddiqi AR *Of Pakistan's First War and Coup*, Defence Journal, Karachi, June-July, 1985.

Siddiqi AR *Mohammad Akbar Khan Talks on Kashmir and Rawalpindi Conspiracy Case*, Defence Journal, Karachi, June-July, 1985.

Sinha SK *Operation Rescue : Military Operations in Jammu and Kashmir, 1947-49*, Vision Books, New Delhi, 1977.

Tinker, Hugh *India and Pakistan, A Political Analysis*, 2nd edition, Pall Mall, London, 1967.

United Nations *Year Book of the United Nations (1947-48), (1948-49), (1949-50) and (1950-51).*

Wolpert, Stanley *Jinnah of Pakistan*, Oxford University Press, New York, 1984.

Ziegler, Philip *Mountbatten : The Official Biography*, Collins, London, 1985.

Index

We have produced ample evidence to show that Hindus of many different castes, even including brahmins, in the sixteenth and early seventeenth centuries travelled freely by sea. This seems to represent a revision of Basham's view, quoted at length earlier, that Hindus almost stopped sea trading some time after the Muslim invasions. C.R. Boxer notes Tamil Hindus still trading in the sixteenth century, "But elsewhere the Hindu merchants remained shore-based, and they shipped their goods in Muslim vessels." This, he says, was a result of newly important caste taboos which meant that by 1500 sea voyages were seen by high caste Hindus as defiling.[94] Simkin also notes "a new, Brahmanical ruling that Hindus sinned by travelling on the sea".[95]

Other recent authors have more nuanced views. Moreland noted vanias and chettyars travelling by sea, while Gokhale said that vanias travelled freely by sea,[96] and Bouchon finds Gujarati vanias less concerned with "Black Water" prohibitions than were Kerala Hindus.[97] Panikkar claims that while it is "perhaps true" that north Indians "had some kind of ceremonial objection to the sea," south Indians never did.[98] Stein quotes south Indian inscriptions of the thirteenth and fourteenth centuries which refer to the spiritually polluting effect of the sea, and says that although sea trade was carried on by Hindus before and after this time, these inscriptions "probably look forward to the general prohibition of sea travel for Hindus in more modern times".[99] Arun Das Gupta says that "the banians too travelled considerable distances by sea apparently disregarding the so-called taboo on sea voyages".[100] Finally, in an earlier piece I claimed that there was no evidence in any contemporary source of caste prohibitions on sea travel, and certainly vanias were not affected by such restrictions. I attributed Muslim dominance at 1500 to conversions and Muslim solidarity.[101]

The authorities clearly differ. What, however, was the theoretical position? Classical and medieval texts quoted by P.V. Kane in sum point to sea travel being prohibited to the top three varnas, or sometimes only to brahmins. In either case, by a process of imitation other castes, including even sudras, also became reluctant to engage in sea travel and trade.

However, sea travel, even for brahmins, seems to have been regarded as undesirable, but not a heinous crime. After all, brahmins and others did earlier travel to Southeast Asia, and various *smritis* refer to regulations for sea trade, and the interest to be paid on trade money.[102]

In the sixteenth and seventeenth century literature there is, my earlier work to the contrary, evidence of prohibitions on sea travel. Thus in Bengal the merchants and weavers of Cassimbazar refused to travel by sea, because "their cast or lineage is such they will lose their birthright if they come upon salt water."[103] When the English at Surat tried to get a Hindu writer (whose name and caste we unfortunately do not have) to go to the Persian Gulf he refused, "it being utterly against the custome of his cast".[104] A century earlier in Malabar Pires noted that "the Nayar's religion forbids him to eat at sea, except by permission of his chief Brahman in case of dire necessity".[105]

There are several points to be made here. First, the consensus is that earlier prohibitions were less draconic, and could be ignored. Later, the arrival of the Muslims, and their technical superiority in nautical matters, helped displace many Hindus. This contributed to an increasingly rigid ban on Hindu travel by sea, as also did the decline of more "rational" Buddhism. Second, in any period there is a clear gap between theory and practice, between, for example, what the *Arthasastra* says a king should do and what he actually does. Nevertheless, it is still necessary to explain now people could reconcile themselves to this. Third, there is of course no one universally accepted Great Tradition attitude to sea travel (or to anything else). There are, clearly, both temporal and regional variations. In particular, the *Bombay Gazetteer's* account of Gujarati brahmins leaves an impression of a rather mixed and not very "pure" regional varna. Many in the late nineteenth century were peasants and cultivators.[106] Perhaps, then, the position of Gujarati brahmins on sea travel, as they proclaimed it to the powerful vania community, was more "lax" than the interpretations of other brahmins in other areas.

Nevertheless, my data does show a theoretical and Great Tradition antipathy to travel by sea, and yet considerable travel, by people of high castes as well as low. How was this possible? The evidence seems to show that if polluted food could be avoided sea travel was acceptable; it was not travel by sea, but the likelihood of impure people on board polluting one's food which was the problem.

A general fear of coming into contact with other, contaminating, people, especially Muslims, has been noted by several authors.[107] Meilink-Roelofsz elaborates, claiming of non-Muslim Gujaratis that "hampered as they were by ritual precepts, they could not themselves embark in Moslem ships for fear of coming into contact with unclean persons." In south India this was not the case as there the ships were Hindu and indeed were restricted to certain castes.[108] But only Bouchon lays adequate stress on the key problem, that of food.[109]

It follows that if supplies of ritually pure food could be ensured, sea voyages were presumably not a problem, though there is no way of knowing exactly how strictly the requirement was observed. If one travelled on an all-vania ship, or an all-Tamil Hindu one there was presumably no problem. However, most ships, as we noted, had communally mixed people on board, and we have excellent contemporary evidence of the importance of food restrictions, and how they were overcome. Linschoten noted that Hindu and Jain vanias would rather starve than eat or drink with foreigners (i.e. people of other castes). Thus when they sailed on Portuguese ships from Goa to Cochin, "they make their provisions for so long time as they thinke to stay upon the way, which they take aboard with them, and thereupon they feede, and if the time falleth out longer, they then made account of their water and provisions beeing all spent, as it hapned when I sailed from Goa to Cochin, they had rather die for hunger and thirst than once to touch and use the food and water of the Christians or of any other. . .[110]

Pyrard noted exactly the same phenomenon : "All, of these Gentiles are badly off when they go a sea voyage, or get into gaol, or among a crowd of strangers : then most put up with dried and preserved fruits, or with a kind of rice, half cooked and then dried, which keeps two or three years. It is greatly used for provisioning all the Indian vessels. . ."[111] Later he described in detail how the Moucois (Mukkuvars of Malabar) went about meeting the food problem when sailing, as they usually would, with people not of their caste or religion. "These Moucois who escorted us make no fires in their Almedie [largish coastal craft], wherefore they cook beforehand large pots of rice, and keep it steeped in cold water, so that it does not go bad. Moreover, they fry a large quantity of salt fish. . .; they also carry a quantity of fruit and cocos : upon this fare they supported themselves, as we did, too".[112]

Hinduism and sea trade were related in other ways. We have noted one group of sailors at sea performing a rite to the god Muthiah to get wind.[113] We also have a description of an interesting rite performed by Konkan Hindus. According to Linschoten, "When they will make a voyage to Sea, they use at the least fourteene days before they enter into their ships, to make so great a noyse with sounding of Trumpets, and to beat pots, that it may be heard and seene both by night and day, the ship being hanged about with flagges, wherewith (they say) they feast their Pagode, that they may have a good Voyage. The like doe they at their returne for a thankesgiving fourteene dayes long. . . ".[114] It appears that these ceremonies were to ensure, and later give thanks for, a safe voyage. One would expect purification ceremonies to be undertaken after a voyage, but so far no evidence of these has been found, at least for this period.

Two final, more speculative, points may be raised in conclusion. Lannoy claims to see a reverse correlation between the strength of "Brahmanic ritualism" and the degree of freedom enjoyed by merchants. When the letter of the law was observed traders were intolerably restricted. The escapes were either conversion to Buddhism or Jainism, or acceptance of anticaste bhakti precepts.[115] This suggestion is at the least interesting. It is surely no coincidence that the Hindu vanias in Gujarat were Vaishnava devotionalists, this being an off-shoot of bhakti.

No doubt Gujarati vanias today are notorious for their orthodoxy, but it is questionable whether this was the case in the sixteenth century, when the efforts of such saints as Vallabhacharya led to a community very much in a state of flux and change. Further to this point of rather slight brahmanic influence on trading communities, it should be remembered that while the vanias are either vaishnavite or Jain, nearly all Gujarati brahmins are shaivite. This division, plus the apparently low purity of many Gujarati brahmins noted above, helps to explain the following nineteenth-century phenomenon, which may apply to the sixteenth century also : "many Vanias show (the brahmins) little honour, and on the whole Gujarat Brahmans meet with less consideration than the Brahmans of other parts of the (Bombay) Presidency".[116] The situation in Tamilnadu with respect to the chettyars may have been similar, but this awaits further investigation.

An alternative explanation is suggested by Milton Singer's concept of "compartmentalization." He notes how modern Madras businessmen appear to behave differently, in terms of religious and caste observance, at home and in the office. His informants told him there was "no conflict because business and religion are separate spheres. . .separate both in the sense of a physical separation of the two spheres and in the sense that different norms of behaviour and belief were appropriate to the two spheres".[117] This notion is certainly suggestive, and appears to be directly related to our problem. It may also be worth considering in other areas, as in helping to explain how Hindus could work closely with Muslim rulers. Even more relevant, R.S. Khare has recently, in his important studies of food, noted how "some of these informants would request me not to disclose in the village how they ate in the city. Such cases of double ethics one for the village and the world outside —was clearly established for frequently travelling men".[118]

To demonstrate the usefulness, if any, of these last two explanations obviously needs much more research. It may be suggested, in conclusion, that all three explanations for Hindu trade by sea in the sixteenth century could be combined. Vanias in particular were less influenced by brahmanical restrictions than were most Hindus. To the extent that they were bothered by sea travel, they were able to reconcile themselves first by doing the best they could to avoid contaminated food at sea, and second by excusing their travel as being necessary for the pursuit of *artha*, compensating for any vestigial bad *karma* by leading ostentatiously devout (but not necessarily orthodox) lives on shore.

NOTES

1. Lucy Carroll deals with some of the modern political implications in her article "The Seavoyage Controversy and the Kayasthas of North India, 1901-1909," *Modern Asian Studies* XIII (1979), pp. 265-99.
2. A. Jan Qaisar, "Nature of Evidence for Shipping in the Persian Sources : A Selective Study," paper written for the seminar "Indian Maritime History, A.D. 1500-1800," Visva Bharati, March 1979.
3. L. Varadarajan, "Traditions of Indigenous Navigation in Gujarat," paper written for the International Conference on Indian Ocean Studies, Perth, Aug. 1979; see also R.M. Mehta, "A Note on some Indian Navigational Terms," *Indica* XVI (1979), pp. 208-11.
4. Auguste Toussaint, *History of the Indian Ocean* (Chicago, 1966), p. 107.
5. Barros, III, i, 7.
6. See Ashin Das Gupta, "Gujarati Merchants and the Red Sea Trade, c. 1700-1725" in Blair B. Kling and M.N. Pearson, eds., *The Age of Partnership : Europeans in Asia before Dominion* (Honolulu, 1979); Ashin Das Gupta, "The Maritime Merchant, c. 1500-1800," Presidential Address, Medieval Section, Indian History Congress, 1974, p. 7; S. Arasaratnam, "Indian Commercial Groups and European Traders, 1600-1800; Changing Relationships in Southeastern India, "*South Asia* I (1978), p. 45; M.A P. Meilink-Roelofsz, "Trade and Islam in the Malay-Indonesian Archipelago Prior to the Arrival of the Europeans," in D.S. Richards, ed., *Islam and the Trade of Asia* (Philadelphia, 1970), pp. 151-2; on pilgrims see Dr. John Fryer, *A New Account of East India and Persia*,3 vols. (London, 1909-15), III, 163. For early eighteenth-century crews see Ashin Das Gupta, *Indian Merchants and the Decline of Surat, c. 1700-1750* (Wiesbaden, 1979), pp. 39-43, and pp. 10-12 for a summary of the pedlars debate. There is a good section on the crews of fifteenth century ships in G.R. Tibbetts, *Arab Navigation in the Indian Ocean before the Coming of the Portuguese* (London, 1971), pp. 58-63. See also Abul Fazl, *Ain-i Akbari*, trans. H. Blochmann, 3 vols. (Delhi, 1978), I, 290-1, for a late sixteenth century list of people on ships, and their salaries.
7. W.H. Moreland, *India at the Death of Akbar* (Delhi, 1974), pp. 198-9. Here and elsewhere Moreland's ideas reflect the unfortunate fact that he did not have access to Tomé Pires' important contemporary account.
8. As examples, G.F. Hudson, "The Medieval Trade of China," in D.S. Richards, ed., p. 166; Genevieve Bouchon, "Les musulmans du Kerala á l'époque de la Découverte portugaise," *Mare Luso-Indicum* II, Geneva, 1973, pp. 33, 45, 50; G.F. Hourani, *Arab Sea-*

faring (Beirut, 1963), p. 83; O.K. Nambiar, *The Kunjalis, Admirals of Calicut* (London, 1963), pp. 2, 14, 19, 28-29; R.B. Serjeant, *The Portuguese off the South Arabian Coast* (Oxford, 1963), p. 3; Vitorino Magalhães-Godinho, *l'Economie de L'empire portugais aux XV et XVI siècles* (Paris, 1969), pp. 558, 565; C.G.F. Simkin, *The Traditional Trade of Asia* (London, 1968), pp. 168-9; Toussaint, pp. 49-57; Bailey W. Diffie and George D. Winius, *Foundation of the Portuguese Empire, 1415-1580* (Minneapolis, 1977), chapters 12 to 15; K.M. Panikkar, *India and the Indian Ocean* (London, 1945), p. 36; D.R. Gadgil, *Origins of the Modern Indian Business Class* (New York, 1959), p. 3. More details on Moreland's views in "The Ships of the Arabian Sea about A.D. 1500," *Journal of the Royal Asiatic Society*, Jan. and April 1939. For details on the technical knowledge of these pre-Portuguese sailors, see A. Teixeira da Mota, "Méthodes de navigation e cartographie nautique dans l'Ocean Indian avant le XVI siècle," *Studia*; no. 11 (Jan. 1-63), pp. 49-91, and G.R. Tibbetts, *Arab Navigation, passim*. For a most interesting comparison with the Chinese, very much to the disadvantage of the Portuguese, see J. Needham, "Abstract of Material Presented to the International Maritime History Commission at Beirut, "in M. Mollat, ed., *Sociétiés et Compagnies de Commerce en Orient et dans l'Océan Indien* (Paris, 1970), pp. 146-54.

9. C.R. Boxer, *The Portuguese Seaborne Empire, 1415-1825* (London, 1969), pp. 44-45; Serjeant, p. 3; Moreland, *Death of Akbar*, pp. 245-6; Bouchon, p. 46; Arun Das Gupta, "The Gujaratis and Europeans in Asian Trade, 1500-1624," in B. De, ed., *Essays in Honour of Professor S.C. Sarkar* (New Delhi, 1976), p. 203; B.G. Gokhale, "Capital Accumulation in XVIIth Century India," *Journal of the Asiatic Society of Bombay*, vols. 39-40, p. 54.
10. Simkin, p. 84; similarly, Boxer, pp. 44-45.
11. A.L. Basham, *The Wonder that was India* (London, 1963), p, 233, and also A.L. Basham, "Notes on Seafaring in Ancient India," *Studies in Indian History and Culture* (Calcutta, 1964), pp. 162-6.
12. Conti's account in R.H. Major, ed., *India in the Fifteenth Century* (London, 1857), p. 26.
13. *The Book of Duarte Barbosa*, 2 vols, (London, 1918-21), II, 64-65.
14. See Barros, I, viii, 1; Zain-ud-din, *Tohfut-ul-Mujahideen*, trans. M.J. Rowlandson (London, 1833), pp. 156-158; Correa, I, 346-7; Couto, VII, x, 17.
15. "Descriçâo das terras da India Oriental . . ."Biblioteca Nacional, Lisbon, f. 1v.
16. Castanheda, I, x.
17. Serjeant, p. 10.
18. "Descriçâo," ff. 15-15v; Barbosa, I, 108-11; *The Suma Oriental of Tomé Pires*, 2 vols. (London, 1944), I, 41-47.
19. See Thevenot's account in S.N. Sen, ed., *Indian Travels of Thevenot and Careri* (New Delhi, 1949), pp. 35, 294; Barros, II, vi, 2; M.R.

Majumdar, *Cultural History of Gujarat* (New York, 1967), p. 136, and especially Jean Aubin, "Albuquerque et les négociations de Cambaye," *Mare Luso-Indicum*, I, Geneva, 1971, pp. 9-12.

20. *Cartas de Affonso de Albuquerque*, 7 vols. (Lisbon, 1884-1935), I, 307.
21. See Pires, I, 41-47; II, 264-5; Correa, II, 253; Braz de Albuquerque, *Commentaries of Afonso Albuquerque*, 4, vols., (London, 1875-84), III, 87-88; Castanheda, III, cxxx; "Descrição," f. 17v.
22. Pires, I, 45.
23. Zain-ud-din, p. 158.
24. Diogo do Couto, *Vida de D. Paulo de Lima Pereira* (Lisbon, 1903), pp. 54-55.
25. See Barros, I, v, 6-7; Castanheda, I, xiii, xvii, xix; "Descrição," ff. 36-37. See especially Barros I, viii, 1 for a general description of the pre-Portuguese spice trade, and Castanheda II, lxxv for details on its focus on Calicut. For the ships see Correa, I, 122-4.
26. "Descrição," ff. 55v-56; A.B. de Bragança Pereira, *Arquivo Português Oriental* (Bastorá, Goa, 1937-40), I, ii, p. 139.
27. Barbosa, II, 79.
28. Pires, I, 82; *Itinerary of Ludovico di Varthema* (London, 1928), p. 61.
29. Pires, I, 81-82. On the Malabar situation, see especially G. Bouchon, op. cit.
30. Barbosa, II, 177.
31. Pires, II, 254-5; Barros, II, vi, 3, 6.
32. Correa, IV, 168.
33. Pires, I, 45.
34. See Jean Aubin, pp. 9-10; Barbosa, I, 147, f.n., 187, f.n. They were noticed by William Finch in the early seventeenth century "speaking another language, and for the most part sea-men." W. Foster, ed., *Early Travels in India* (London, 1921), p. 135. For their origins, ethnography, and disappearance from Rander see *Gazetteer of the Bombay Presidency*, vol. IX part 2, *Gujarat Population*; *Musalmans and Paris* (Bombay, 1899), pp. 14-15; *Gujarat State Gazetteers, Surat District* (Ahmedabad, 1962), p. 946.
35. Boxer, p. 57; António de Abreu, "Orçamento do Estado da India," in *Studia* IV (1959), p. 221; DI, IV, 340; APOCR, IV, 26; see also J.H. van Linschoten, *The Voyage of John Huyghen van Linschoten to the East Indies*, 2 vols. (London, 1885), I, 266-67.
36. See APOCR, III, 384-6; V, 407; *Boletim do Instituto Vasco da Gama* XXVI, 32-33; Barros, IV, iv, 12; DUP, I, 402. For Hindu sailors in the early eighteenth century see Ashin Das Gupta, *Indian Merchants*, pp. 40-41.
37. "Termo das Fianças, HAG, I, 142-2v.
38. Pires, I, 82; Nambiar, 83, 89; Linschoten, I, 282; A.C. Roy, *A History of Mughal Navy and Naval Warfare* (Calcutta, 1972), p. 10; Ralph Fitch in W. Foster, ed., op. cit., p. 45.

39. Barbosa, II, 64-65; *Travels of Franciso Pyrard de Laval* 2 vols. (London, 1887-90), I, 386, 444.
40. See Nambiar; K.M. Panikkar, *Malabar and the Portuguese* (Bombay, 1929); and chapter two.
41. See their works cited in footnotes 8 and 14 above.
42. See chapter two for piracy.
43. Couto, *Lima Pereira*, pp. 54-55.
44. Barbosa, I, 186 and f.n.; Couto, X, v, 7; X, vi, 4.
45. APOCR, VI, 745-6.
46. Linschoten, II, 255.
47. "Colecção de São Lourenço," ANTT, III, ff. 53v, 54, 60, 65, 94.
48. *Studia*, 3, pp. 75-76.
49. "Termo das Fianças,,' HAG, I, ff. 7-7v.
50. For example, APOCR, I, ii, pp. 121-2; *Diário de 3rd Conde de Linhares* (Lisbon, 1937-43), p. 243; Fazenda IX, 247v-248v; J.I. de Abranches Garcia, ed., *Archivo da Relação de Goa*, 2 vols. (Nova Goa, 1872-74), pp. 704-6.
51. Fazenda, VI, 131v-2; AHU, "India," caixa 16, opinion of 1644; APOCR, VI, 1267-73; LM, XIV, f. 319; Assentos, II, 247, Pyrard (II, 246) noted how on the ships of the Cambay convoy "they have all their own ensigns, and the livery of their several lords on their flags; and the merchants to whom they are consigned recognize them afar off."
52. "Chancellaria, D. Affonso VI," ANTT, XXXIX, f. 372-2v. Descriptions of vanias in Goa in Linschoten, I, 228, 252-5; in Bassein in DI, X, 19-20, letter of 1575.
53. Fazenda, V, 99-99v; "Correspondencia de Baçaim," HAG, I, 88.
54. "Consultas do serviço de partes," "HAG", III, 151v.
55. Couto, X, iv, 8.
56. Pyrard, II, 250.
57. "Alvarás e provisoes de sua Magestade," HAG, II, 16-20v; "Chancellaria, D. João IV," ANTT, XIX, 236v.
58. Pyrard, I, 364-5.
59. Das Gupta, in *The Age of Partnership*, p. 133.
60. Jean Aubin, "Le royaume d' Ormuz au début de XVI siècle," *Mare-Luso Indicum*, II, 1973, especially pp. 95-6, 150-1, 162-4.
61. Pires, I, 81.
62. G. Bouchon, pp. 13-14.
63. *Gazetteer of the Bombay Presidency*, vol. IX, part 1, *Gujarat Population : Hindus* (Bombay, 1901), p. 18.
64. S.D. Goitein, *Studies in Islamic History and Institutions* (Leiden, 1966), pp. 338, 349-50. It is true that this evidence comes from an earlier period, but Prof. Das Gupta assures me that vania trade to the Red Sea continued after this.
65. "Descrição," f. 22.

66. *Alguns documentos do archivo nacional da Torre do Tombo* (Lisbon, 1892), p. 453; Castanheda, IV, xvi; V, ix; Barros, III, i, 10; III, ii, 3.
67. Barros. IV, iv, 8.
68. Varthema, pp. 37-38; Artur Viégas, ed., *Relação Anual das coisas que fizeram os Padres da Compánhia de Jesus nas suas missões*, 3 vols. (Lisbon, 1930-42), by Padre Fernão Guerreiro, II, 389.
69. See on this destruction Barros, IV, iv, 8.
70. DI, III, 712; IV, 263, 754.
71. Couto, VII, ix, 14.
72. ANTTCC, 1-78-8; 3-15-31; 3-16-21; "Cartas de Vice Reis," ANTT, nos 75 and 146.
73. LM, XXVIII B, ff, 422-6; "Livro Vermelho," HAG, II, 90; AHU, "India," caixa 6, report of 1619; APOCR, III, 681-3; "Fianças," HAG, II, 143-4; A Botelho da Costa Veiga, ed., *Relação das plantas e descrições de todas as fortalezas, cidades, e povoações que os Portugueses tem no Estado da India Oriental* (Lisbon, 1936), p. 20; Assentos, II, 74; Bragança Pereira, ed., IV, 2, 1 (Bocarro's "Livro"); ANTTDR, LI, 131-iv.
74. ANTTDR, LI, 131-1v; 136-6v.
75. Viégas, ed., II, 389-93.
76. See footnote 64.
77. Bragança Pereira, IV, 1, pp. 113-4; "Arrematação das rendas do estado," HAG, I, ff. 211-214v.
78. See Das Gupta, in *The Age of Partnership, passim.*
79. APOCR, V, 957.
80. Ibid., III, 680-3.
81. "Ordens Régias," HAG, I, 191; AHU, "India," caixa 14, petition of 1642; W. Foster, ed., *The English Factories in India*, 17 vols. (London, 1904-54), 1642-45, p. 161; 1646-50, p. 144; Violet Barbour quotes a contemporary mention of "Hindus and other foreign peoples" at the Amsterdam bóurse in the seventeenth century. Violet Barbour, *Capitalism in Amsterdam in the Seventeenth Century* (Baltimore, 1950), p. 57. How unfortunate we have no names to help identify such long distance travellers.
82. Niels Steensgaard, *The Asian Trade Revolution of the Seventeenth Century* (Chicago, 1974), p. 91.
83. *Factories*, 1622-23, pp. 264, 298, 324, 329; 1624-29, pp. 71-72.
84. *Factories*, 1630-33, pp. 284, 299.
85. See Das Gupta in *The Age of Partnership*, pp. 132-5, for eighteenth century examples.
86. *Letters Received by the East India Company from its Servants in the East, 1602-1617*, 6 vols. (London, 1896-1902), V, 239; C.D. Ley, *Portuguese Voyages* (London, 1947), p. 344.
87. See Abd-er-Razzak in Major, ed., pp. 5-6; DI, I, 504-5; AHU "India," caixa 3, no. 82.
88. "Livro Morato," HAG, ff. 331-1v; Fazenda, II, 31v.

89. Jean Aubin, "Le royaume d'Ormuz au début de XVI siècle," *Mare Luso-Indicum* II, 1973, *passim*.
90. For example, Serjeant, pp. 32-34, 70; *Factories*, 1624-29, pp. 70-72.
91. "Cartas e Ordens," HAG, III, f. 73; Jourdain, pp. 95, 99, 104; *Letters Received*, 1602, 13, pp. 51, 167, 181, 227; *Recueil des voyages qui ont servi á l'établissement et aux progrez de la Compagnie des Indes Orientales*, 12 vols. (Rouen, 1725), VII, 462.
92. The standard account of these people is in the works of Ashin Das Gupta, cited in footnote 6 above.
93. M.K. Gandhi, *An Autobiography* (Ahmedabad, 1940), pp. 83-84.
94. C.R. Boxer, pp. 44-45.
95. C.G.F. Simkin, p. 84.
96. B.G. Gokhale, "Capital Accumulation," p. 54; W.H. Moreland *India at the Death of Akbar*, p. 246.
97. Bouchon, *Mare Luso-Indicum*, II, 46.
98. Panikkar, *India and the Indian Ocean*, p. 29.
99. Burton Stein, "Coromandel Trade in Medieval India," in John Parker, ed., *Merchants and Scholars* (Minneapolis, 1965), p. 52.
100. Arun Das Gupta in De, ed., p. 203.
101. M.N. Pearson, *Merchants and Rulers in Gujarat* (Berkeley, 1976), pp. 14-15.
102. P.V. Kane, *History of Dharmasastra*, 4 vols. (Poona, 1930-73), I, 28; III, 610, 933-8; IV, 393; Bouchon, 11-15; R.K. Mookerji, *A History of Indian Shipping* (Bombay, 1957), pp. 41-43, A.L. Basham, "Notes," pp. 162-6.
103. Quoted in Gokhale, p. 54.
104. *Factories*, 1624 29, p. 163.
105. Pires, I, 81.
106. *Gazetteer of the Bombay Presidency*, vol. IX, part 1, *Gujarat Population : Hindus* (Bombay, 1901), pp, 1-54.
107. Boxer, p. 45, for example.
108. M.A.P. Meilink-Roelofsz, *Asian Trade and European Influence in the Indonesian Archipelago between 1500 and about 1630* (The Hague, 1962), pp. 63-64.
109. Bouchon, pp. 11-15.
110. Linschoten, II, 255.
111. Pyrard, I, 379. In Gujarati this sort of rice is called "paua.'
112. Ibid., I, 427. It will be remembered that when Gandhi made his first voyage to England in 1888, he was unsure of the ship's food and "I therefore never took meals at table but always had them in my cabin, and they consisted principally of sweets and fruits which I had brought with me." (*Autobiography*, p. 31).
113. Conti in Major, ed., p. 26.
114. Linschoten, II, 227-8.
115. Richard Lannoy, *The Speaking Tree : A Study of Indian Culture and Society* (London, 1971), pp. 232-6.
116. *Bombay Gazetteer*, op. cit. in footnote 106, p. 22.

117. Milton Singer, *When a Great Tradition Modernizes* (New York, 1972), pp. 320-1.

118. R.S. Khare, *The Hindu Hearth and Home* (New Delhi, 1976), p. 13, and in more detail pp. 243-63. It must however be noted that Khare finds this conflict between ideological and pragmatic views to be only "apparent." According to him, there is "internal consistency at the ideological level" : R.S. Khare, *Culture and Reality* (Simla, 1976), pp. 161-2. There remains a question of how many real Indians perceive this "consistency". At an everyday behavioural level, his informants clearly do find a discrepancy.

Appendix

DETAILS OF SOME PASSES GRANTED BY THE PORTUGUESE IN 1618-22

All spellings are as in the original

Serial No.	Nakhoda of Ship	Owner of Ship	Size of Cargo (Candis)
1.	Fraute	King of Caixem	700
2.	Hamar Bene Aguer	King of Caixem	300
3.	Sumbol	King of Caixem	200
4.	Xeque Ali	King of Bijapur	1500
5.	Assengi	King of Bijapur	4000
6.	Aligi Paricar	King of Bijapur	1000
7.	Fraqui Ali	King of Bijapur	1500
8.	Mamedegi	King of Bijapur	2000
9.	Srimamo qtroli (?)	King of Bijapur	3000
10.	Mahamedina (Persian)	Jahangir	3000
11.	Marguean, casta Mouro	King of Caixem	350
12.	Frahate, Mouro	King of Caixem	100
13.	Frahate, Mouro	King of Caixem	450
14.	Margean Abdulla	King of Caixem	100
15.	Ambar	King of Caixem	750
16.	Amede, Mouro	King of Caixem	50
17.	Sidi Belele	King of Ahmadnagar	600
18.	Sidi Surar, Abexim	King of Ahmadnagar	2500
19.	Sidi Surar, Abexim (No. 2)	King of Ahmadnagar	?
20.	Mamede Carabas, Persian	King of Ahmadnagar	1500
21.	Cutti Poera	King of Cananor	1000
22.	Cottu Muxa	Sandegaro Chaty of Barcelor	800
23.	Faquiali	King of Bijapur	2000
24.	Xequi Aly	King of Bijapur	4000
25.	Coje Mohedim	King of Bijapur	1500
26.	Oensa Nagi	King of Bijapur	4000

27.	Mahamedina	Jahangir	3000
28.	Usem, casta Naitia	King of Kanara	300
29.	Coje Mahamede	King of Bijapur	500
30.	Joar	King of Caixem	350
31.	Mahamede Hagi	King of Bijapur	300
32.	Hasengi Branga	King of Bijapur	1000
33.	Fatealy	King of Bijapur	1500
34.	Asigi Paricar	King of Bijapur	700
35.	Mahamede Agi	King of Bijapur	200
36.	Mahamedina	Jahangir	3000
37.	Cide Johar, Abexim	King of Ahmadnagar	1000
38.	Sidi Surur, Abexim	King of Ahmadnagar	2000
39.	Mamedagy, naitea de nacao	King of Ahmadnagar	300
40.	Faqui Aly	King of Bijapur	1500
41.	Hansanagi boura	King of Bijapur	1000
42.	Alagi Paricar	King of Bijapur	700
43.	Edula Nacardina	King of Bijapur	700
44.	Mocdina	King of Bijapur	300
45.	Ambar	King of Caixem	500
46.	Johar	King of Caixem	600
47.	Rehana	King of Caixem	500
48.	Sada Auahit	King of Caixem	250
49.	Coje hamede, Persian	King of Ahmadnagar	2500
50.	Mamede	King of Bijapur	1000
51.	Coje hamede, Persian	King of Ahmadnagar	1500
52.	Melique	King of Bijapur	2000
53.	Melique Vafade	King of Bijapur	1000
54.	Melique Asal	King of Bijapur	2000
55.	Melique Muratta	King of Bijapur	4000
56.	Esmael	King of Bijapur	1000
57.	Melique Aza	King of Bijapur	1000
58.	Melique Fatta	King of Bijapur	4000
59.	Mahamed	King of Bijapur	300
60.	Mameda da casta Naitea	King of Ahmadnagar	1000
61.	Hagi Raza Vany, Persian	King of Ahmadnagar	700
62.	Nassurdy Mahamed, Persian	Jahangir	2000
63.	Melique Cazal	King of Bijapur	2000
64.	Melique Murata	King of Bijapur	4000
65.	Melique Vafa	King of Bijapur	1000
66.	Ismael	King of Bijapur	1000
67.	Melique Fatu	King of Bijapur	4000
68.	Melique Aza	King of Bijapur	1000

69.	Mamede	King of Ahmadnagar	2000
70.	Mamede Carabaca, Persian	?	1500
71.	Coje Nizamo, Persian	Jahangir's mother	2500
72.	Said Aluiguifi	King of Caixem	100
73.	Alixaca	King of Caixem	250
74.	Auarid	King of Caixem	700
75.	Lalegi Dossy	Antonio Leitao of Diu	600
76.	Auay	King of Caixem	500
77.	Agi Mamede	Aminal Malucco, a captain of the King of Bijapur	400

Note : one candil is equal to approximately 500 lbs.
Source : HAG, "Consultas," III, 44-136v.

Index